BORN TO BE GOOD

BORN TO BE GOOD

The Science of a Meaningful Life

Dacher Keltner

W. W. NORTON & COMPANY   *New York* · *London*

For information about permission to reproduce selections from this book,
write to Permissions, W. W. Norton & Company, Inc.,
500 Fifth Avenue, New York, NY 10110

For information about special discounts for bulk purchases, please contact
W. W. Norton Special Sales at specialsales@wwnorton.com or 800-233-4830

Manufacturing by RR Donnelley, Harrisonburg
Book design by Iris Weinstein
Production manager: Julia Druskin

Library of Congress Cataloging-in-Publication Data

Keltner, Dacher.
Born to be good : the science of a meaningful life / Dacher Keltner.
— 1st ed.
p. cm.
Includes bibliographical references and index.
ISBN 978-0-393-06512-1 (hardcover)
1. Interpersonal relations. 2. Helping behavior. 3. Altruism.
4. Cooperation. I. Title.
HM1106.K45 2009
155.2'32—dc22
2008042492

W. W. Norton & Company, Inc.
500 Fifth Avenue, New York, N.Y. 10110
www.wwnorton.com

W. W. Norton & Company Ltd.
Castle House, 75/76 Wells Street, London W1T 3QT

1 2 3 4 5 6 7 8 9 0

TO THOSE WHO HAVE GIVEN ME SO MUCH *JEN*:
THE FAMILY WHO RAISED ME—JEANIE KELTNER, MY MOTHER;
RICHARD KELTNER, MY FATHER; AND ROLF KELTNER, MY BROTHER—
AND THE FAMILY WHO SUSTAINS ME—
MOLLIE MCNEIL, MY WIFE; AND NATALIE AND SERAFINA KELTNER-
MCNEIL, MY DAUGHTERS.

CONTENTS

PREFACE

SOME SCIENTIFIC INSIGHTS arise in powerful, fleeting experiences—a startling observation, a dream, a gut feeling, a sudden realization. My own thinking about human emotion has taken a longer course, reflecting the intersection of the long arc of life and the scientific data I have gathered.

I have been led to the idea that emotion is the source of the meaningful life. My mother, an English professor and student of Romanticism, and my father, an artist guided by Lao Tzu and Zen, cultivated in me the conviction that our best attempts at the good life are found in bursts of passion captured in plot turns in the prose on a page or oil layered on a canvas. This idea proved to have the deepest scientific promise in the hands of Charles Darwin, who believed that brief emotional expressions offer clues to the deep origins of our design, and Paul Ekman, who figured out how to bring quantifiable order to the thousands of movements of the face.

Born to Be Good is the product of my family and scientific upbringings, and attempts answers to three age-old questions. The first is: How can we be happy? Legions of recent empirical studies on happiness have led to best-selling books. We have learned about the difficulties of knowing what makes us happy (Dan Gilbert's *Stumbling on Happiness*), the importance of optimism (Martin Seligman's *Authentic Happiness*), and that, for most of us, relationships are the surest route to happiness, and seeking happiness through financial gain is an illusion (Jonathan Haidt's *The Happiness Hypothesis*).

Born to Be Good offers a new answer to the question of how we can be happy. In honor of my mother and father, you might call it the Zen Romanticism thesis. The idea is that we have evolved a set of emotions that enable us to lead the meaningful life, emotions such as gratitude, mirth, awe, and compassion (the Romanticism thesis). The key to happiness is to let these emotions arise, to see them fully in oneself and in others, and to train the eye and mind in that practice (the Zen thesis).

Born to Be Good engages a second old question: What are the deep origins of our capacity for kindness? We are witnessing a renewed debate about our origins. Advances in DNA measurement, in archeology, and in the study of our primate relatives are yielding striking new insights into the history of humanity, where we came from, how we dispersed, how we evolved. Embedded in these discoveries is an answer to the question of where our capacity for goodness comes from. *Born to Be Good* reveals how survival of the kindest may be just as fitting a description of our origins as survival of the fittest.

Finally, *Born to Be Good* asks: How can we be good? We are in a period of probing moral reflection. U.S. children rank twentieth of twenty-one industrialized countries in terms of social well-being. The moral stature of the United States has fallen dramatically in the past eight years. Deep concerns about genocide, inequality, and global warming raise doubts about whether a hopeful future for the human race is justified on any grounds. There is a hunger for new views of the nature and practice of human goodness. Just look at the crowds that attend every talk by the Dalai Lama.

Born to Be Good reveals a straightforward answer to the question of how to be good: rely on emotions like amusement, gratitude, and compassion to bring the good to others to completion. To give life to this idea, *Born to Be Good* offers a conversation between Darwinian views of the origins of human goodness (you'll be surprised to learn that Darwin believed that sympathy is our strongest passion) and gems from the great traditions of East Asian and Western thought.

The road map of *Born to be Good* follows the development of my own thinking about answers to these questions. It begins with a discussion, in chapter 1, of the Confucian concept of *jen*, which refers to kindness, humanity, and reverence. I introduce the concept of the *jen* ratio, a simple but powerful way of looking upon the relative balance of good and uplifting versus bad and cynical in life. The *jen* ratio honors my interests in Eastern philosophy and in parsimonious measurement. It is a way to think about the clues to happy marriages, well-adapted children, healthy communities and cultures.

The next three chapters take the reader on a tour of the latest discoveries in evolutionary approaches to emotion. We start in chapter 2 by considering Darwin's nuanced analyses of the many positive emotions. Contrary to what many may assume, Darwin believed that these emotions were the basis of our moral instinct and capacity for good. Darwin and Confucius would have been very content collaborators.

From Darwin we travel to New Guinea, and Paul Ekman's paradigm-shifting studies of the universality of facial expression. As a result of the empirical science that followed this work, we have arrived at three new ideas about emotion that are summarized in chapter 3: Emotions are signs of our commitment to others; emotions are encoded into our bodies and brains; emotions are our moral gut, the source of our most important moral intuitions.

In chapter 4 we look back in time to glean what has been learned about the evolution of human goodness. This ever-changing evolutionary science provides a context for understanding the origins of the positive emotions, where the smile comes from, why we are wired to trust and to care. The chapter brings together insights from the study of our close primate relatives, from archeology, and from hunter-gatherer cultures. The reader may be surprised to learn that:

- We are a caretaking species. The profound vulnerability of our offspring rearranged our social organization as well as our nervous system.
- We are a face-to-face species. We are remarkable in our capacity to empathize, to mimic, to mirror.
- Our power hierarchies differ from those of other species; power goes to the most emotionally intelligent.
- We reconcile our conflicts rather than fleeing or killing; we have evolved powerful capacities to forgive.

- We live in complex patterns of fragile monogamy, preferring monogamy but often showing patterns of serial monogamy.

Each of the remaining eight chapters is devoted to the science of different emotions that give rise to high *jen* ratios. This science is rooted in Darwin's deepest insight about human emotion: that the visible expressions of emotion that we observe today are clues to the ancient behaviors that led some mammals to fare well in the tasks of survival, reproduction, and care of offspring. This science would not exist without the methodology Paul Ekman and Wallace Friesen gave to the field: the ability to measure millisecond movements of muscles in the face. Every chapter tries to honor the insights about emotion found in art, literature, and philosophy.

In chapter 5, I begin where I began, with a serendipitous discovery of the embarrassment display. I show how embarrassment acts as an appeasement display, prompting others to forgive and forget. The smile is revealed, in chapter 6, to have evolved as a signal of equality and trust, and as a sign of the life well lived. Chapter 7 examines how laughter evolved as a unique signal of play and levity, detailing the varieties of laughter and how laughter promotes healthy responses to trauma. In chapter 8, I examine the much-maligned act of teasing. I show, building on the study of fools and jesters as well as the philosophy of language, how teasing is actually a remarkable act of pretense and drama, and enables people to negotiate conflicts and hierarchies. In chapter 9, I survey the startling new science of touch: it makes people trust, it increases body weight in premature babies and reduces depression in adults in nursing homes, it builds strong immune systems. In our lab we have documented that people can communicate compassion, love, and gratitude to a stranger with one-second touches to the forearm. In chapter 10 I reveal lasting insights about humans' reproductive relations, profiling the new discoveries on oxytocin, a neuropeptide that promotes devotion, and how it is released during nonverbal displays of love. Compassion is the focus of chapter 11. Darwin thought it was the foundation of our moral sense and cooperative communities. I focus on new discoveries about a branch of the nervous system, the vagus nerve, which is centrally involved in compassion. I conclude, in chapter 12, by examining awe. I begin by talking about John Muir's experiences of awe in the Sierras that led to the environmentalist movement and trace back to revolutionary thinkers in the West who transformed our experience of awe, from a religious experience to something that can be felt in nature, toward others in art, and in spiritual experience. I then rely on studies of goosebumps, dinosaurs, and beauty to tell a story about the evolution of this fascinating emotion, and how it enables us to fold into cooperative social collectives.

In carrying out the science and writing that led to *Born to Be Good*, I have become more acutely aware of our capacity for *jen*. I see it in the smiles of friends, our many modest ways, melodious laughter, moving touches, and the readiness with which we care, appreciate, and revere. Seeing these capacities in our species has brought a bit of the good in me to completion. I hope the same will prove to be true for you.

ACKNOWLEDGMENTS

I HAVE SO MANY to acknowledge and appreciate.

I would like to thank my mentors Phoebe Ellsworth, Robert Levenson, and Paul Ekman for initiating me into the scientific study of emotion, when the field was just getting off the ground. I have learned so much from my collaborator colleagues: George Bonanno, Lisa Capps, Serena Chen, Avshalom Caspi, James Gross, Deborah Gruenfeld, Jonathan Haidt, Oliver John, Ken Locke, Robert Knight, Ann Kring, David Matsumoto, Terrie Moffitt, Michael Morris, Terrie Moffitt, and Gerben van Kleef. In writing two textbooks, I have been made wiser in perspective and prose by Tom Gilovich, Jennifer Jenkins, Richard Nisbett, and Keith Oatley. In friendly conversations, many of the ideas here have been enlivened and brought into focus by Chris Boas, Nathan Brostrom, Gustave Carlson, Christine Carter, Claire Ferrari, Michael Lewis, Jason Marsh, Peter Platt, and especially Tom Gilovich, Leif Hass, Mollie McNeil, and Frank Sulloway, who dwelled in the ideas here in ways that let me view their promises and pitfalls through different eyes.

I thank those who have funded my research: The National Institute of Mental Health, the Russell Sage Foundation, the Templeton Foundation, the Fetzer Foundation, the Metanexus Institute, the Mind and Life Institute, and the Positive Psychology Network. I am deeply grateful to Tom and Ruth Ann Hornaday, founders of Berkeley's Greater Good Science Center, a place to cultivate the next generation of scientists interested in how we are born to be good, and Lani and Herb Alpert, for their support in building our magazine, *Greater Good*.

This book would not exist were it not for the astonishing empirical science of my students: Cameron Anderson, Jennifer Beer, Brenda Buswell, Belinda Campos, Adam Cohen, David Ebenbach, Jennifer Goetz, Gian Gonzaga, June Gruber, Erin Heerey, Matthew Hertenstein, Elizabeth Horberg, Emily Impett, Michael Kraus, Carrie Langner, Jennifer Lerner, Alexander LuoKogan, Lorraine Martinez, Chris Oveis, Paul Piff, Sarina Rodrigues, Laura Saslow, Lani Shiota, Emiliana Simon-Thomas, John Tauer, Ilmo van der Löwe, Kris Vasquez, and Randall Young.

I was guided at the outset in far-reaching ways by my agent, Linda Lowenthal. And my dear editor, Maria Guarnaschelli, early on found the soul of this book, wouldn't let me waver, and brought the good in this book to completion.

Duchenne smiles and grateful touches to you all.

BORN TO BE GOOD

1

Jen Science

ANTON VAN LEEUWENHOEK changed how we look at the natural world. Born in Delft, the Netherlands, in 1632, he came from a family of brewers and basket-makers. Van Leeuwenhoek was peacefully settled into his life as a fabric maker, minor city official, and wine assayer until he started grinding up lenses to build simple microscopes to get a better look at the drapes in his shop. His curiosity led him to place the algae of nearby lakes under his three- to four-inch single-lens microscopes, as well as the cells of fish, his sperm, and the plaque of two old men who had never cleaned their teeth. He was the first to study bacteria, blood cells, and spermatozoa. He opened humanity's eyes to the microbiological world, changing our understanding of who we are.

This book offers a Darwinian lens onto a new science of positive emotion. We'll call this new science *jen* science, in honor of the Confucian concept of *jen*. *Jen* is the central idea in the teachings of Confucius, and refers to a complex mixture of kindness, humanity, and respect that transpires between people. Alienated by the violence, the materialism, and the hierarchical religion of his sixth- and fifth-century BC China, Confucius taught a new way of finding the meaningful life through the cultivation of *jen*. A person of *jen*, Confucius observes, "wishing to establish his own character, also establishes the character of others." A person of *jen* "brings the good things of others to completion and does not bring the bad things of others to completion." *Jen* is felt in that deeply satisfying moment when you bring out the goodness in others.

Jen science is based on its own microscopic observations of things not closely examined before. Most centrally, it is founded on the study of emotions such as compassion, gratitude, awe, embarrassment, and amusement, emotions that transpire between people, bringing the good in each other to completion. *Jen* science has examined new human languages under its microscope—movements of muscles in the face that signal devotion, patterns of touch that signal appreciation, playful tones of the voice that transform conflicts. It brings into focus new substances that we are made of, neurotransmitters as well as regions of our nervous system that promote trust, caring, devotion, forgiveness, and play. It reveals a new way of thinking about the evolution of human goodness, which requires revision of longstanding assumptions that we are solely wired to maximize desire, to compete, and to be vigilant to what is bad.

Seeing the world through this Darwinian lens of *jen* science could very well shift your *jen* ratio. The *jen* ratio is a lens onto the balance of good and bad in your life. In the denominator of the *jen* ratio place recent actions in which someone has brought the bad in others to completion—the aggressive driver who flips you off as he roars past, the disdainful diner in a pricey restaurant who sneers at less well-heeled passersby. Above this, in the numerator of the ratio, tally up the actions that bring the good in others to completion—a kind hand on your back in a crowded subway car, the young child who compliments the elderly woman on her bathing suit as she nervously dips her toe in a swimming pool, the woman who laughs as a stranger accidentally steps on her foot. As the value of your *jen* ratio rises, so too does the humanity of your world.

Let's give the *jen* ratio a little life. An after-school moment at my daughters' playground yields the following. In the numerator: two boys laugh, giving each other noogies on the head, girls do handstands and cartwheels, giggling at their butt-thumping mistakes, in the soft expanse of a grassy field, kids dog pile on a young boy deliriously clasping the football to his chest. In the denominator: a boy teases a smaller boy about his shoes; two girls whisper about another girl who tries to enter into their game of unicorn. This minute of playground life yields a *jen* ratio of 3/2, or 1.5. A pretty good scene. In an interminable, eight-minute line to buy stamps I see 24 varieties of exasperation, from sighs to glares to threatening groans, and one guy laughs three times. 3/24 = .125.

One can apply the *jen* ratio to any realm: our interior life, more satisfying and more trying periods of a marriage, the tenor of a family reunion, the goodwill of a neighborhood, the rhetoric of presidents, the spirit of historical eras. Think of the *jen* ratio as a lens through which you might take stock of your attempt at living a meaningful life.

THE *JEN* RATIO AND THE HEALTH OF NATIONS

Simple measures can yield powerful diagnoses. Apgar scores, blood pressure indexes, emotional intelligence quotients all take minutes to derive but reveal the course a life can take. What measure would you propose to diagnose the social well-being of our times? Murder rates? The GDP? The distribution of wealth to those on the top compared to those at the bottom? The percentage of citizens who believe in the resurrection? The speed with which people laugh at Homer Simpson? If I were given one metric to take the temperature of the social well-being of the individual, the marriage, a school, community, or culture, the *jen* ratio would be my choice.

For the individual, new studies are finding that a high *jen* ratio, a devotion to bringing the good in others to completion, is the path to the meaningful life. Engaging in five acts of kindness a week—donating blood, buying a friend a sundae, giving money to someone in need—elevates personal well-being in lasting ways. Spending twenty dollars on someone else (or giving it to charity) leads to greater boosts in happiness than spending that money on oneself (even though most people think that spending money on themselves would be the surer route to happiness). When pitted against one another in competitive economic games, cooperators and those who forgive selfish partners fare better than competitors in terms of economic outcome. New neuroscience suggests we are wired for *jen*: When we give to others, or act cooperatively, reward centers of the brain (such as the nucleus accumbens, a region dense with dopamine receptors) hum with activity. Giving may enhance self-interest more than receiving.

What works for individuals works in marriage: Bringing the good of a romantic partner to completion (and not the bad) yields many positive returns. One of the most toxic developments in marriage is the emergence of a low *jen* ratio. In over twenty studies of how romantic partners explain one another's actions, the couples heading toward divorce routinely attribute the good things in their intimate life to the selfish motivations of their partner ("he brought me flowers just to butter me up for his golf weekend"). Regrettably, they just as readily pin the responsibility of their hassles, struggles, and crises on their partners ("if she'd clean the back seats of our car every now and then we wouldn't have things growing under there"). Happier couples are guided by a high *jen* ratio: They generously give credit to their partner and see hidden virtues accompanying their partner's foibles and faults.

And what's true of individuals and marriages is true of nations: Nations whose citizens bring the good in others to completion thrive. High *jen* ratios are proving to be a hallmark of healthy societies. In 1996, Paul Zak and colleagues asked random samples of participants in various countries to answer the following question: "Generally speaking, would you say that most people can be trusted, or that you cannot be too careful in dealing with people?" (See figure below for levels of trust across cultures.) After statistically controlling for appropriate variables, such as economic development, Zak and colleagues found that for every 15 percent increase in the trust of a nation's citizens, their economic fortunes rise by $430. Trust facilitates economic exchange with fewer transaction costs, including fewer failed negotiations, adversarial settlements, and needless lawsuits. With increased trust among a citizenry, discrimination and economic inequality fall. High *jen* ratios promote a society's economic and ethical progress.

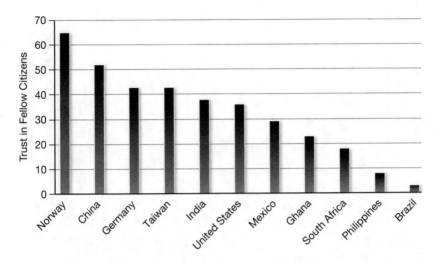

One cannot help but be struck by the cultural variation in the levels of trust presented in this figure. As a generalization, Zak found that Scandinavian and East Asian cultures are more trusting than South American and Eastern European cultures. Poorer nations (India) are often more trusting than wealthier nations (the United States).

JEN YEN

In the recent explosion of studies on social well-being, signs of a loss of *jen* in the United States are incontrovertible. The percentage of Americans who trust their fellow citizens has dropped 15 percentage points in the past fifteen years. Many indicators of our culture's poor health—increasing feelings of anomie, greater loneliness, the trend toward less happy marriages—are on the rise. U.S. adults now have one-third fewer close friends in their circle of intimates than twenty years ago. Young babies have more physical contact with their Hummer-like baby strollers than from the touch of their parents' hands. In a recent UNICEF study of twenty-one industrialized nations, U.S. children ranked twentieth in terms of overall well-being (see figure below).

The decline in our social well-being has been blamed on the abandonment of the classics of Western civilization in higher education, moral relativism, and the loss of religious faith. Others have floated different causes of the wearing down of the social fabric: the

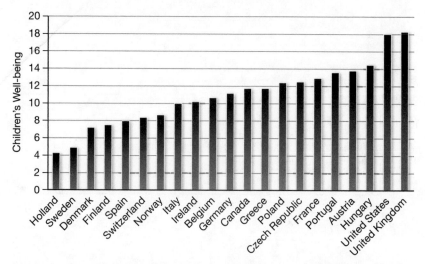

The measure of childhood well-being was based on the sum of six measures: material well-being, health and safety, education, peer and family relationships, behaviors and risks, and children's own subjective well-being. Lower scores indicate higher overall rankings reflective of greater childhood well-being in the country.

disappearance of modesty, precocious sexuality, the proliferation of mediated communication, and fast food.

I see these disheartening social trends as the culmination of a broader ideology about human nature, one with a *jen* ratio trending toward 0. This ideology has influential advocates from Sigmund Freud to evolutionary theorists. The strongest proponents of this view are found in the halls of economics departments. Their characterization of human nature, known widely as rational choice theory, extends to evolutionary thought, psychological science, and the field of emotion—my own disciplines. It has blinded us to the science and practice of *jen*.

At the center of this theory is *Homo economicus*, the latest stage of human evolution, a hominid first portrayed by philosophers, Adam Smith the most well-known, who were awestruck by the great expansion of wealth, technology, and progress during the Industrial Revolution. First and foremost, *Homo economicus* is selfish. Every action of *Homo economicus* is designed to maximize self-interest, in the form of experienced pleasure, advances in material wealth, or, in evolutionist thought, the propagation of genes.

When "pleasure centers" were first discovered in 1954 in a region of the limbic brain known as the septum, rats, neither hungry nor thirsty, would press bars for hours on end just to have that region stimulated. Economists assume that you and I have much in common with those rats, that we are wired for the perpetual pursuit of personal gratification. And now a new field, neuroeconomics, is beginning to back this up: Of the sixty miles of neural wiring of the human brain, the regions involved in representing the basic rewards, such as sweet tastes or pleasant scents, also light up like Christmas lights in fMRI scans at the prospect of winning money.

If humans are wired to maximize the fulfillment of desire, a second claim readily follows: Competition is a natural and normative state of affairs. Competition subjects our unbounded self-interest to the rational order of the marketplace in which supply and demand constrain the fulfillment of our desires. Cooperation and kindness are, by implication, cultural conventions or deceptive acts masking deeper self-interest. As a case in point, consider the debate about generous acts toward strangers. These acts cost you in terms of time, energy, and missed opportunities and, most irrationally, benefit genetically unrelated individuals. Such acts fall outside the reach of stalwart evolutionary concepts of inclusive fitness (the generous act benefits those individuals who share our genes) and reciprocal altruism (the generous act is eventually reciprocated and thus enhances the individual's welfare). The conclusion: These generous acts are evolutionary "misfires" or "strategic errors"; they are misapplications of more self-interested systems such as the love of kin or friends who reciprocate.

Evolving in societies of selfish, competitive gratification machines gave rise to a third design feature of *Homo economicus*: a mind wired to prioritize the bad—the toxic food, hidden snake, cuckolding friend, backstabbing peer—over the good. That the bad is stronger than the good is evident in several findings. Our memory for bad things is more robust than for good things (it's a wonder any of us ever go back to a high-school reunion). Economic losses loom larger than their equivalent gains: Losing twenty dollars stings more than finding twenty dollars pleases. Slides of negative stimuli (pictures of mutilated faces or a dead cat) trigger stronger activation in brain regions associated with evaluative judgments than slides of positive stimuli (pictures of a pizza slice or a bowl of chocolate). Negative things, Paul Rozin observers, contaminate the positive more readily than vice versa. A cockroach walks near a plate of truffles and they're ruined. Place a truffle near a pile of cockroaches and . . . they're still disgusting.

These claims, that humans are wired to pursue self-interest, to compete, and to be vigilant to the bad rather than the good, are no doubt familiar to you. In fact, they lie at the heart of the intellectual traditions that have shaped Western thought. These core assumptions guide the thinking of founding figures in fields such as psychoanalysis, economics, political theory, and evolutionary theory:

The very emphasis of the commandment "Thou shalt not kill" makes it certain that we are descended from an endlessly long chain of generations of murderers, whose love of murder was in their blood as it is perhaps also in ours.　　　**SIGMUND FREUD**

Objectivist ethics, in essence, hold that man exists for his own sake, that the pursuit of his own happiness is his highest moral purpose, that he must not sacrifice himself to others, nor sacrifice others to himself.

AYN RAND

Of mankind we may say in general they are fickle, hypocritical, and greedy of gain.　　　**MACHIAVELLI**

The natural world is "grossly immoral." . . . Natural selection "can honestly be described as a process for maximizing short-sighted selfishness."

GEORGE C. WILLIAMS

These deeply ingrained assumptions tilt the *jen* ratio of our culture toward extremely low values, and they have led us astray in our pursuit of the meaningful life.

THE MIDLIFE CRISIS OF *HOMO ECONOMICUS*

Social life would be a strange affair if humans had no sense of self-interest, no competitive drive, no attunement to the bad in the world, if their *jen* ratios had no input into the denominator. Just ask the parents of children with Williams syndrome. These children often have elfin features and intellectual deficits (for example, language delays, distractibility, difficulties controlling attention). Yet they can possess a fascinating combination of savantlike talents: They may sparkle in conversation, be exquisitely sensitive to music and loud sounds, and be preternaturally at ease with others. Unsettling from the parents' perspective is that children suffering from Williams syndrome often have little sense of individuality, few boundaries, and a pure and at times problematic interest in the welfare of others. They tend to catapult themselves into friendly relations with people they have just met—the pimply checkout guy, the traffic officer, the psychotic panhandler hustling a dollar reciting poetry about the end of the world.

Clearly we are wired to pursue self-interest, to compete, and to be vigilant to the bad. Those tendencies make evolutionary sense, they are built into our genes and nervous systems. They are part of human nature. But that is half the story. Several lines of thought suggest that *Homo economicus*, and its proponents, are experiencing a crisis in faith, that this model of human nature is only part of our story.

Economists themselves have asked whether maximizing self-interest is always the core motive of human action. This questioning found galvanizing expression in a 1988 book, *Passions Within Reason*, by Cornell economist Robert Frank. Frank offers a series of observations that reveal humans to be more than bar-pressing rats seeking that pleasure-center buzz. Here's one: From the perspective of maximizing personal desire, why do we tip a waitress after grabbing breakfast in a diner in a town we will never set foot in again? There really is no personal gain to this costly act, no promise of reciprocity, no speeded-up service the next time we frequent the restaurant, no burnishing of our reputation in the eyes of the other diners who witness our largesse. Our economic lives, Frank observes, are punctuated by actions that harm our self-interest while enhancing the welfare of others: generosity toward co-workers, acts of charity to far-away children and the protection of other species, buying Girl Scout cookies at exorbitant prices. The health of long-term bonds, Frank continues, depends upon high *jen* emotions like compassion, grati-tude, and love.

An outpouring of empirical findings has lent credence to Frank's prescient claims. Consider one such set of results, generated by studies that use the ultimatum game. In the ultimatum game, an allocator is given a sum of money, say $10, and told to keep a cer-tain amount and allocate the rest to a second participant, a respon-der, who can either accept or reject the offer. If rejection is the choice, neither player receives anything. The enlightened self-interested approach would be for the allocator to give the respon-der 1¢, or perhaps in a fit of generosity $1, and for the responder to accept. In the end, both individuals' overall wealth is advanced, and Adam Smith smiles.

Across ten studies of people in 12 different cultures, however, economists Ernst Fehr and Klaus Schmidt found that 71 percent of the allocators offered the responder between 40 and 50 percent of the money. Most people showed a strong preference for near-equality. Remember these are strangers they're making these offers to, not kith and kin. People around the world will sacrifice the enhancement of self-interest in the service of other principles: equality, a more favorable reputation, or even, God forbid, the advancement of others' welfare.

So we don't always act in the pure pursuit of self-interest. What about a more basic question: Does material gain make us happy? Certainly this is a pervasive notion. Seventy-four percent of today's undergraduates cite economic gain as the primary motive for going to college (compared to 25 percent twenty years ago, above other motives, such as developing a philosophy of the meaningful life or contributing to the greater good). We are increasingly defining our needs in terms of gratifying materialistic desire. Look at the table below, adapted from Alain de Botton's *Status Anxiety*, which por-trays recent shifts in the percentage of U.S. citizens who consider different products to be basic necessities.

	1970	2000
SECOND CAR	20	59
SECOND TV	3	45
MORE THAN ONE TELEPHONE	2	78
CAR AIR CONDITIONING	11	65
HOME AIR CONDITIONING	22	70
DISHWASHER	8	44

Does money make us happy? The answer for those who have very little is yes. Material gain allows individuals in the lowest economic strata to avoid the innumerable problems associated with economic deprivation, including depression, anxiety, compromised resistance to disease, and higher mortality rates.

For those in the middle classes and above, however, the association between money and happiness is weak or nonexistent. Researchers have now asked millions of people the simple question: "How satisfied are you with your life right now?" It is not personal wealth, the strength of the stock market, inflation, or fluctuations in interest rates that cause the ebb and flow in our personal well-being. This same literature reveals time and time again that what makes us happy is the quality of our romantic bonds, the health of our families, the time we spend with good friends, the connections we feel to communities. When our *jen* ratios are high in our close relations, so are we.

THE *JEN* OF BLUSHES, LAUGHS, SMILES, AND TOUCH

When I began my study of emotion fifteen years ago, the bad is stronger than the good thesis found a comfortable home in the literature on emotion. Empirical studies were quick to find that there are more words in the English language that represent negative than positive emotions. Only one signal of positive emotion in the face, the smile, had been studied, compared to the five or six signals of negative emotion. Nothing was known about how positive emotions activate our autonomic nervous system, which controls basic bodily functions like digestion, blood flow, breathing, and sexual response. These empirical facts led many in the field to the view that positive emotions are in reality by-products of negative states. For example, my first response is fear toward someone I don't know, and upon recognizing that such a person is familiar and safe, I experience warmth and love; affection is in actuality the cessation of fear. Scientists were quick to take the next step: The negative emotions are rooted deeper in human nature than positive emotions, and are a more active currency in our daily living. The *jen* ratio of the science of emotion hovered near 0.

As I have applied the Darwinian lens to the emotions the past fifteen years, another swath of human nature was revealed to me, positive emotions that bring the good in others to completion. I kept encountering fleeting two- or three-second emotions that belonged in the numerator of the *jen* ratio. As I painstakingly coded hundreds of hours of people talking about the deaths of their spouses, I witnessed laughs that acted like brief journeys to a more peaceful state. In the surge of affection between romantic partners, I saw brief head tilts, smiles, and open-handed gestures. I saw teasers, once playfully vicious, soothe their targets with quarter-second touches to the shoulders and fleeting eye-to-eye affirmation. I observed sexual interest sweep over flirtatious young suitors' faces in lip puckers and lip bites. After being startled, people recovered their poise briefly, but then collapsed, disheveled, into a state of embarrassment, looking askance, blushing, touching their faces, and smiling awkwardly. When I presented images of these abashed participants to others, these viewers would sigh vocalizations that provided clues to the origins of compassion.

The canonical studies of human emotion, studies of the universality of facial expression, of how emotion is registered in the nervous system, how emotion shapes judgment and decision making, had never looked into these states. The groundbreaking studies of emotion had only examined one state covered by the term "*happiness*." But research is often misled by "ordinary" language, the language we speak rather than the language of scientific theory. Happiness is a diffuse term. It masks important distinctions between emotions such as gratitude, awe, contentment, pride, love, compassion and desire—the focus of this book—as well as expressive behaviors such as teasing, touch, and laughter. This narrow concentration on "happiness" has stunted our scientific understanding of the emotions that move people toward higher *jen* ratios. By solely asking "Am I happy?" we miss out on the many nuances of the meaningful life.

My hope is to shift what goes into the numerator of your *jen* ratio, to bring into sharper focus the millisecond manifestations of human goodness. I hope that you will see human behavior in a new light, the subtle cues of embarrassment, playful vocalizations, the visceral feelings of compassion, the sense of gratitude in another's touch to your shoulder, that have been shaped by the seven million years of hominid evolution and that bring the good in others to completion. In our pursuit of happiness we have lost sight of these essential emotions. Our everyday conversations about happiness are filled with references to sensory pleasure—delicious Australian wines, comfortable hotel beds, body tone produced by our exercise regimens. What is missing is the language and practice of emotions like compassion, gratitude, amusement, and wonder. My hope is to tilt your *jen* ratio to what the poet Percy Shelley describes as the great secret of morals: "the identification of ourselves with the beautiful which exists in thought, action, or person, not our own." The key to this quest resides in the study of emotions long ignored by affective science. It will require that we return to Darwin's country home, Down House, in Kent, England, and that we travel with Paul Ekman to the highlands of New Guinea.

2

Darwin's Joys

I N 1967, PAUL EKMAN was lucky to land his Cessna on a small clearing in a jungle of New Guinea (the plane had lost a wheel on takeoff). He arrived with a packet of photos, film equipment, and a hypothesis in hand. He was there thanks to U.S. government malfeasance in South America. A government agency had been funding counterinsurgency research in left-leaning South American countries under the guise of public opinion research. When a congressional committee found out, the funds were quickly shunted to a promising but relatively unknown young researcher—Ekman—for a cross-cultural study of emotion recognition.

Ekman traveled to New Guinea to ascertain whether people from a culture remote from Western influence, a people living in pre-industrial, hunter-gatherer conditions, would interpret photos of facial expressions of six emotions—anger, disgust, fear, happiness, sadness, and surprise—as you or I would. Ekman doubted he would obtain such findings. He was steeped in the cultural relativist assumptions of the era. He had little notion that this work, however it turned out, would catalyze the study of emotion. Nor did he suspect that this work would unravel time-honored notions about the place of emotion in human nature. He was making the trip in the discovery-oriented spirit of Charles Darwin.

DARWIN'S JOYS

Darwin's *Expression of the Emotions in Man and Animals* sold 9,000 copies in its first printing, becoming a best seller in its day. *Expression* sparked spirited discussion among scientists and laypeople alike in the parlors and cafés of Victorian England. Perhaps most important to Darwin, the book met with modest smiles of approval from his wife, Emma. She rightly anticipated that a book on emotional expression would unsettle prim and hierarchical Victorian ideology less than Darwin's *On the Origin of Species*.

In the aftermath of *On the Origin of Species*, Darwin responded to a barrage of attacks on his theory of evolution. The most impassioned of these attacks centered upon whether natural selection could account for the design of human beings. Creationists were giving way to evolutionary accounts of the origins of rocks, reefs, shales, mollusks, barnacles, and finches. Their empirical open-mindedness had clear limits, though: They recoiled at the possibility that humans themselves were products of evolution, descended from apes, shaped by natural selection, not touched by the hand of God or designed according to ideas of perfectibility.

Emotions have long been a battleground for competing views of human nature. This once again proved to be the case in this clash between evolutionists and creationists. Creationists like the anatomist Sir Charles Bell argued that God had graced humans with special facial muscles that allowed them to express uniquely human emotions, lofty, "higher" moral sentiments like sympathy, shame, or rapture, emotions unknown to "lower" species. The uniqueness of human facial expression, by implication, was proof of the discontinuity of the human species from other species. The subtle emotional expressions you might observe in your spouse or children, Bell reasoned, were the visible traces of the handiwork of God. Those facial muscles were part of a rationale for why humans should be at the top of the great chain of being, master of other species.

With his astounding powers of observation, Darwin took on this challenge. He marshaled an eclectic variety of data to document "mental continuities" between human and animal expression. Amid the fifteen to twenty-five letters that arrived each day at his house from his correspondents (Darwin was a prolific letter writer), observations flowed in about emotional displays in other animals, expressive outbursts that sounded remarkably akin to those of our loved ones. There were detailed accounts of terriers frowning in concentration, pug dogs mugging intelligence, and monkeys throwing temper tantrums. Darwin himself catalogued the emotional displays of his dog Polly, a terrier who lay curled in devotion at Darwin's feet as he labored on his articles, books, and correspondences in his study. Darwin closely studied his beloved ten children. He relied on the astute observations of mothers he knew to limn the expressions of emotion—wails of pain, laughter and smiling, sulking and tenderness—that emerged early in life. He turned with fervor to a new technology—photography. Darwin collected over 100 photos, photos of actors portraying different emotions and of the specific muscle actions produced by electrical stimulation.

These kinds of data led Darwin to a rich portrayal, the most detailed ever achieved, of human emotional expression. Unlike the scientists who would study emotion some one hundred years later, Darwin posited many different positive emotions, sixteen by my count. His theorizing about the evolution of positive emotions revealed a *jen* ratio unusually high for our field. The table below summarizes Darwin's observations, which are a poetic periodic chart of how emotions are expressed, the embodied signatures of brief subjective states.

DARWIN'S DESCRIPTIONS OF EMOTIONAL EXPRESSION

EMOTION	EXPRESSIVE BEHAVIORS
NEGATIVE EMOTION	
ANGER	TREMBLE, NOSTRILS RAISED, MOUTH COMPRESSED, FURROWED BROW, HEAD ERECT, CHEST EXPANDED, ARMS RIGID BY SIDES, EYES WIDE OPEN, STAMP GROUND, BODY SWAYS BACKWARDS/FORWARDS
ANXIETY	INNER ENDS OF EYEBROWS RAISED, CORNERS MOUTH DEPRESSED

CONFUSION	STAMMER, GRIMACES, TWITCHINGS OF FACIAL MUSCLES
CONTEMPT	LIP PROTRUSION, NOSE WRINKLE, EXPIRATION, PARTIAL CLOSURE OF EYELIDS, TURN AWAY EYES, NOSE WRINKLE, UPPER LIP RAISED, SNORT, EXPIRATION
DISAGREEMENT	CLOSE EYES, TURN AWAY FACE
DISGUST	LOWER LIP TURNED DOWN, UPPER LIP RAISED, EXPIRATION, MOUTH OPEN, SPITTING, BLOWING OUT PROTRUDING LIPS, CLEAR THROAT SOUND, LOWER LIP, TONGUE PROTRUDED
EMBARRASSMENT	LITTLE COUGH, BLUSH
FEAR	TREMBLE, EYES OPEN, MOUTH OPEN, LIPS RETRACTED, EYEBROWS RAISED, CROUCH, PALE, PERSPIRATION, HAIR STANDS ON END, MUSCLES SHIVER, YAWN
GRIEF	INNER ENDS OF EYEBROWS RAISED, CORNERS OF MOUTH DEPRESSED, FRANTIC MOVEMENTS, MOTIONLESS, HEAD HANGS, EYELIDS DROOP, ROCK TO AND FRO, FACE PALE, MUSCLES FLACCID, EYELIDS DROOP, CONTRACTED CHEST, TEARS, OBLIQUE EYEBROWS, DEEP SIGHS, BEAT HANDS, EYELIDS DROP
GUILT	GAZE AVERSION, SHIFTY EYES, GRIMACE
HORROR	BODY TURNED AWAY, SHRINKING, ARMS PROTRUDING, SHOULDERS RAISED, ARMS PRESSED AGAINST CHEST, SHUDDER, DEEP INSPIRATION/EXPIRATION, SHUT EYES, SHAKE HEAD
INDIGNANCE, DEFIANCE	FROWN, BODY ERECT, HEAD ERECT, SQUARE SHOULDERS, CLENCH FISTS
ILL TEMPER	FURROWED BROW, NOSE WRINKLE, LIP CORNERS PULLED DOWN
NEGATION	HEAD THRUST BACK
OBSTINATENESS	MOUTH FIRMLY CLOSED, LOWERED BROW, SLIGHT FROWN
PAIN	WRITHE ABOUT, PIERCING CRIES, GROANS, LIPS COMPRESSED, RETRACTED, TEETH CLENCHED, WILD STARE, PERSPIRATION, FURROWED BROW, NOSTRILS DILATED, PROFUSE SWEATING, PALLOR, UTTER PROSTRATION, EYES CLOSED, SQUARE MOUTH (LIPS CONTRACTED), COMPRESSION OF EYEBALL, MUSCLE AROUND EYES

	CONTRACTED, PYRAMIDAL MUSCLE CONTRACTS, UPPER LIP RAISED, NOSTRILS NARROWED, SCALP/FACE/EYES REDDENED, INSPIRATION, SOBBING, LACHRYMAL GLAND SQUEEZED, LAUGHTER, TEARS
PERPLEXED	SCRATCH HEAD, RUB EYES
RAGE	UNCOVERED TEETH, HAIR BRISTLE, FACE REDDENS, CHEST HEAVES, NOSTRILS DILATED, QUIVER, TREMBLE, TEETH CLENCHED, RESPIRATION LABORED, FRANTIC GESTURES, VEINS ON FOREHEAD/NECK DISTENDED, BODY ERECT, BENT FORWARD, ROLL ON GROUND AND KICKING, SCREAMING (CHILDREN), FURROWED BROW, GLARE, PROTRUDING LIPS, RETRACTED LIPS, TOSS ARMS ABOUT, SHAKE FIST, HISSING
RESIGNATION	OPEN HANDS, ONE OVER OTHER, ON LOWER PART OF BODY
SADNESS	CORNER MOUTH DEPRESSED, INNER CORNER EYEBROWS RAISED
SNEER, SNARL	CORNER OF LIP OVER TEETH RAISED
SHAME	BLUSH, HEAD AVERTED, HEAD DOWN, EYES WAVERING, EYES DOWN/AWAY, TURN BODY AWAY, FACE AWAY, BLINKING EYELIDS, TEARS
SULKY	POUT, PROTRUDE LIPS, FROWNING, LIFT SHOULDER AND JERK IT AWAY
TERROR (INTENSE FEAR)	PALLOR, NOSTRILS FLARE, GASPING, GULPING, PROTRUDING EYEBALLS, PUPILS DILATED, HANDS CLENCHED/OPENED, ARMS PROTRUDED, SWEAT, PROSTRATION, BODY RELAXED, EYEBROW CORNERS TIGHTENED AND RAISED, UPPER EYELIDS RAISED, LIP CORNERS PULLED SIDEWAYS, BRISTLING OF HAIR, LOSS OF HAIR COLOR
WEAKNESS, IMPOTENCE APOLOGY	SHOULDER SHRUG, INWARD TURN OF ELBOWS, HANDS EXTENDED WITH PALMS OPEN, EYEBROWS RAISED

POSITIVE EMOTION

ADMIRATION	EYES OPENED, EYEBROWS RAISED, EYES BRIGHT, SMILE
AFFIRMATION	NOD HEAD, OPEN EYES WIDELY
ASTONISHMENT	EYES OPEN, MOUTH OPEN, EYEBROWS RAISED, HANDS PLACED OVER MOUTH

CONTEMPLATION	FROWN, WRINKLE SKIN UNDER LOWER EYELIDS, EYES DIVERGENT, HEAD DROOPS, HANDS TO FOREHEAD, MOUTH, OR CHIN, THUMB/INDEX FINGER TO LIP
DETERMINATION	FIRMLY CLOSED MOUTH, ARMS FOLDED ACROSS BREAST, SHOULDERS RAISED
DEVOTION (REVERENCE)	FACE UPWARDS, EYELIDS UPTURNED, FAINTING, PUPILS UPWARDS AND INWARDS, HUMBLING KNEELING POSTURE, HANDS UPTURNED
HAPPINESS	EYES SPARKLE, SKIN UNDER EYES WRINKLED, MOUTH DRAWN BACK AT CORNERS
HIGH SPIRITS, CHEERFULNESS	SMILE, BODY ERECT, HEAD UPRIGHT, EYES OPEN, EYEBROWS RAISED, EYELIDS RAISED, NOSTRILS RAISED, EATING GESTURES (RUBBING BELLY), AIR SUCK, LIP SMACKS
JOY	MUSCLE TREMBLE, PURPOSELESS MOVEMENTS, LAUGHTER, CLAPPING HANDS, JUMPING, DANCING ABOUT, STAMPING, CHUCKLE/GIGGLE, SMILE, MUSCLE AROUND EYES CONTRACTED, UPPER LIP RAISED
LAUGHTER	TEARS, DEEP INSPIRATION, CONTRACTION OF CHEST, SHAKING OF BODY, HEAD NODS TO AND FRO, LOWER JAW QUIVERS UP/DOWN, LIP CORNERS DRAWN BACKWARD, HEAD THROWN BACKWARD, SHAKES, HEAD/FACE RED, MUSCLE AROUND EYES CONTRACTED, LIP PRESS/BITE
LOVE	BEAMING EYES, SMILING CHEEKS, TOUCH, GENTLE SMILE, PROTRUDING LIPS (IN CHIMPS), KISSING, NOSE RUBS
MATERNAL LOVE	TOUCH, GENTLE SMILE, TENDER EYES
PRIDE	HEAD, BODY ERECT, LOOK DOWN ON OTHERS
ROMANTIC LOVE	BREATHING HURRIED, FACE FLUSHED
SURPRISE	EYEBROWS RAISED, MOUTH OPEN, EYES OPEN, LIPS PROTRUDED, EXPIRATION, BLOWING/HISSING, OPEN HANDS HIGH ABOVE HEAD, PALMS TOWARD PERSON WITH STRAIGHTENED FINGERS, ARMS BACKWARDS
TENDERNESS (SYMPATHY)	TEARS

Here amid Darwin's precise observations, one learns that we cough when embarrassed. Darwin notes subtle distinctions in the displays of admiration and devotion. He reveals that we close our eyes when describing horrors and raise our eyebrows when remembering. When feeling resigned, we rest one open hand upon another on a lower part of our bodies. In a burst of high spirits we just might get caught rubbing our bellies or smacking our lips. Images of friends and family burst into our minds when we read Darwin's descriptions: Confusion is a stammer, grimace, and twitching of facial muscles (my colleagues at faculty meetings); defiance is expressed in the frown, erect body and head, square shoulders, and clenched fists (my daughters when asked to leave a play date).

Why do our emotional expressions look as they do? Why does anger, for example, have the furrowed brow, upper eyelid raise, and tightened, clenched mouth? Why does it not involve any of the thousands of other possible facial muscle combinations? To answer this question, Darwin invoked three principles of expressive behavior. According to the principle of serviceable habits, expressive behaviors are vestiges of more complete actions that have led to rewarding outcomes in our evolutionary history. As a result, they tend to re-occur over time and become reliable signals of internal states and likely actions. Disgust, for example, looks as it does with wrinkled nose, flared nostrils, open mouth and protruding tongue because it is the vestige of vomiting, and signals our experience of revulsion when noxious substances enter the mouth or are at risk of doing so (or noxious ideas risk contaminating the mind). The facial expressions we observe today are a rich shorthand for communicating the possibility of more full-bodied actions—attack, flight, embrace.

Darwin arrived at his second principle of expressive behavior—the principle of antithesis—in part from observations of his stable dog Bob. One of Bob's characteristic displays was the "hot house face," a sullen canine display of drooping head, ears, and tail. Darwin reliably observed this display when Bob was denied pleasure—a run with Darwin in the country, for example. Bob's display, which charmed Darwin so, took the opposite form of the upright ears,

This ape is in the midst of a classic dominance display that will involve postural expansion, expansive arm movements, and the throwing of branches, rocks, and plants in the vicinity. What is also clearly visible is piloerection, a physiological response that leads to the contraction of the muscles surrounding hair follicles and, in humans, is associated with awe, as we shall see. This display is organized around a simple principle of dominance—expand physical size.

head, and tail seen when Bob merrily ran alongside his owner. Here Darwin discerned a broader principle organizing this endearing display of disappointment: the principle of antithesis, which holds that opposing states will be associated with opposing expressions. One of the clearest signs of dominance, shown by alpha apes, CEOs, and pedantic professors alike, is the arms and head akimbo. In this display the individual expands the chest, holds clasped arms behind the head, and leans back. This signal of dominance is the diametrical opposite of the signs of weakness and impotence (see table)—head movements down, shoulder constriction.

Finally, in good Victorian fashion, Darwin held that certain expressive behaviors were organized according to the principle of nervous discharge. This principle holds that excess, undirected energy is released in random expressions, such as head scratches, face touches, leg jiggles, nose tugs, hair twists, and the like. One prevailing metaphor of emotion, at the very heart of Freud's theory of emotional conflict and the psychodynamic mind, is that emotions are like fluids in containers. We boil over, blow our top, get steamed, and feel ready to explode during numerous states, from anger to rapturous ecstasy to sexual desire. Many emotional states, therefore, should produce seemingly random behaviors that reflect the intrapsychic hydraulics of emotion. We tug at our hair when nervous, shake our head when embarrassed, and bite our lips when feeling desire and the impulse to hop in the sack with our dinner date.

This extraordinary culling, sifting, and winnowing of observations of humans and nonhumans left Darwin exhausted and in physical pain at the end of each day of writing, but he turned quickly to his *Expression* book. As he parsed the realm of expressive behavior, tracing it back to our primate predecessors, he realized the critical data that he lacked: a study that would address whether facial expressions are universal to a human species shaped by a common history of selection pressures. He queried English missionaries in other countries (receiving thirty-six responses) about whether they had observed expressions not seen in Victorian England. They had not. Of course, Darwin's manner of asking the question may have encouraged the answers he sought. He returned to his notes about his encounters with the indigenous peoples of Tierra del Fuego, Tahiti, and New Zealand during his five-year voyage on the *Beagle*. When Darwin met the Fuegians, who greeted the passengers of the disembarking *Beagle* naked, with arms flailing and long hair streaming, Darwin was the first to make friends with them by reciprocating their friendly chest slaps. Perhaps in those recollections he saw signs of universal human expression. The definitive data, however, would come 100 years later, in the paradigm-shifting research of Paul Ekman.

This Fuegian, presumably greeting the members of the *Beagle* for the first time, is showing a complex balance of strength and warmth. The right elbow is positioned at an angle, out from the body, expressing subtle traces of the arms akimbo—a dominance display. At the same time the left arm is clasped close to the heart, perhaps a sign of friendship. Note that the dog to the Feugian's left is coiled and ready to strike. Darwin would later write about the affectionate behavior of dogs and cats, which, according to the principle of antithesis, takes the opposite form of this dog's aggressive stance, as evident in the upward and arched motion of the torso and legs.

IN THE HIGHLANDS OF NEW GUINEA

Paul Ekman put Darwin's universality thesis to a simple empirical test. The results of this study provoke controversy, *ad hominem* critique, and sneers at happy-hour conversations at scientific conferences to this day. First, Ekman and his colleague Wallace Friesen took photos of collaborators in Ekman's lab and actors from the local community, posing the facial muscle configurations of six different emotions—anger, disgust, fear, happiness (a broad smile involving the twinkle of the eyes), sadness, and surprise—according to Darwin's detailed descriptions (see pictures below). In a first wave of studies, Ekman and Friesen then asked individuals from Japan, Brazil, Argentina, Chile, and the United States to choose the word, from six (anger, disgust, fear, happiness, sadness, and surprise), that best matched the emotion shown in each photo.

The data gathered in this study would pit two radically different conceptions of emotion against one another (see table below).

Anger   Disgust   Fear

Sadness   Surprise   Happiness

A SUMMARY OF CONSTRUCTIVIST AND EVOLUTIONARY
APPROACHES TO EMOTIONS

QUESTION	CONSTRUCTIVIST APPROACH	EVOLUTIONARY APPROACH
What is an emotion?	Language, beliefs, concepts	Physiological processes in the body
Are emotions universal?	No	Yes
What are the origins of emotions?	Values, institutions, social practices	Natural selection

An evolutionary approach took shape as Ekman started to publish the findings from this first study. The prevailing view of the day—the social constructivist view—emerged out of the influential writings of anthropologists, such as Franz Boas and Margaret Mead. These authors had pioneered thinking about cultural relativism, and the endless variability and moral equivalences of different cultures. Within this tradition, emotions are thought of as social constructions, put together in culturally specific ways according to historically situated values, institutions, practices, and rituals. Emotions at their core are concepts, words, and ideas that shape, and are shaped by, discourse practices such as storytelling, poetry, public shaming, or gossip. What about the expression of emotion across cultures—the question that put Ekman on that wobbly plane to New Guinea? Here the constructivist prediction is that the expression of emotion is analogous in origin, form, and predicted cultural variability to spoken language. Cultures select particular phonemes from the dozens of phonemes the human vocal apparatus can produce to express different concepts in words. The same could be true of emotional expression. Members of cultures, the reasoning held, select different muscle movements to express different emotions. The end result is a prediction of endless cultural variability in the meaning of emotional expression.

The observations, mostly anecdotal, in support of this constructivist view were persuasive. The Inuit were never observed to express anger, even in the most frustrating and unjust circumstances, as when their precious canoes were badly damaged by careless mainland tourists. Upon receiving the news of their husbands dying—nobly—in battle, the wives of seventeenth-century Japanese samurai were observed to smile with pride and love.

In Ekman's first study, individuals from highly modernized cultures demonstrated considerable agreement in their interpretations of the six kinds of facial expressions. The problem, though—quite obvious in the clarity and comfort of hindsight—is that individuals from all of these cultures had been extensively exposed to Western media. Perhaps in those encounters with Hollywood emotion—John Wayne and Doris Day movies, *Howdy Doody* and *Get Smart*

Paul Ekman during a trip to New Guinea.

reruns—Ekman's participants in different cultures had learned how to interpret the facial expressions that he had presented.

As a result, Ekman voyaged to Papua New Guinea. There he lived for several months with a hill tribe from the Foré (pronounced foray) language group that lived in hunter-gatherer conditions. After receiving the blessings of a witch doctor, Ekman recruited nearly 5 percent of the tribe to participate in his study. The Foré who participated in Ekman's study had seen no movies or magazines, they did not speak English or pidgin, they had not lived in Western settlements, and they had not worked for Westerners. Given this history, it would be hard to argue how Western concepts could have penetrated the Foré mind to influence how they would interpret the photos Ekman was to present to them.

In the critical study, Ekman used a judgment method known as the Dashiell method, because the preliterate Foré participants were not well practiced in answering multiple-choice questions. Ekman presented participants with a story appropriate for each of the six emotions. For example, the story for sadness was: "the person's child had died, and he felt sad." Upon hearing the story, Foré participants, both adults and children, selected the emotional expression, from three different ones presented in photos, that best matched the story. If the Foré participants were simply guessing, one would have expected correct identifications of the facial expressions 33 percent of the time—a result that would have conformed to the predictions of social constructivists and their claims about the cross-cultural variation in emotional expression. Foré adults and children, in contrast, were correct 80 to 90 percent of the time in interpreting the six facial expressions—a finding that would have triggered the smile, raised eyebrows, and bright eyes of admiration in Darwin. Untouched by industrialization and modernity, the Foré interpreted those six facial expressions as you or I would.

ACTION UNITS AND THE OBJECTIVE SUBJECTIVE

When Ekman returned to the United States and first presented his results at a conference in anthropology, he was shouted down from the dais. Ideological accusations rang out in the auditorium. Ekman's New Guinea data suggested that a biological facet of emotion—the movement of different facial muscles—was universal. Clearly such a notion is at odds with the constructivist claim that biology plays little role in emotion. Perhaps the chorus of critiques arose because Ekman's data may have been reminiscent of the claims of Social Darwinism—that racial differences are rooted in evolution and biology. Early constructivists like Boas and Mead had soundly routed these Social Darwinist claims (the irony, of course, is that Ekman's data highlight the deep similarities—presumably shaped by evolution—of people from radically different cultures).

Constructivists countered with the most well-cited study of emotion during this era, one that seductively argued that emotional experiences arise out of interpretations prompted by the particulars of the social context and not any specific physiological response. Perhaps the most definitive demonstration of the constructivist thesis would be to show that the same physiological response could lead to radically different emotions given how people interpret the situation they are in.

Such was the aim of Stanley Schachter and Jerome Singer, the authors of this study. In the study, participants were led to believe that the experiment was examining the effects of a vitamin compound, Suproxin, on vision. Most germane to our present interests are the participants who received a shot of epinephrine (adrenaline), which increases blood pressure, heart rate, and the sweatiness of the palms, and who were not told of the effects of the shot. These aroused participants then found themselves in one of two contexts, which unfolded according to theatrical outbursts of a confederate (in cahoots with the experimenter) that would have made the Marx brothers proud. In a euphoria condition, the confederate, sitting across from the heart-palpitating participant in a small room, first crumpled sheets of paper up and attempted jump shots into the trash can. After announcing "I feel like a kid again," he made a paper airplane and launched it into the air. He shot pieces of paper with a rubber-band slingshot, built a tower out of manila folders, and began shimmying with hula hoops left behind a portable blackboard.

In an anger condition a much different emotional drama took place. The confederate and the participant somberly completed the same five-page questionnaire. After questions about his childhood diseases, his father's annual income, and psychiatric symptoms family members have presented, the confederate exploded. When asked how often he had sexual intercourse each week, and "With how many men (other than your father) has your mother had extramarital relationships?" (for which the lowest response category was "4 and under"), the confederate stomped out of the lab room, muttering about the idiocy of the study. Critical to the constructivists' cause, those aroused participants in the euphoria condition reported being much happier than those in the anger condition. A similar physiological response—elevated fight/flight physiology produced by the epinephrine shot—could lead to radically different emotions depending on the interpretation prompted by the particular context. Constructivists around the world cheered.

This study undermined the very foundation of what would become the evolutionary approach to emotion—that the emotions are embodied in distinct, genetically encoded physiological processes universal to humans and shaped by our evolutionary past. Instead, it would seem that emotions can arise out of any physiological response, depending on the interpretation of that experience. The specificity of emotion—whether we experience shame, love, anger, or compassion—and the very nature of emotional experience are the products of culturally based constructive processes taking place in the rich associative networks of the mind.

To counter this ingenious study and its many implications, Ekman confronted a career-imperiling problem: how to measure emotions objectively. What sort of measure could be relied upon to capture fleeting emotional experiences as they stream by in our affective lives? Ideally, this measure could be captured as close to the experience as possible, and used in labs around the world. The most obvious answer is to ask participants to describe their experiences with words, as Schachter and Singer had done. Perhaps the most miraculous expressions of emotion are through words, as in this poem of love by E. E. Cummings:

which is the very
(in sad this havingest
world) most merry
most fair most rare
—the livingest givingest
girl on this whirlingest
earth?
        why you're
by far the darlingest

who (on this busily
nowhere rollingest
it)'s the dizzily
he most him
—the climbingly fallingest
fool in this trickiest
if?
        why i'm
by much the luckiest

what of the wonder
(beingest growingest)
over all under
all hate all fear
—all perfectly dyingest
my and foreverless
thy?
        why our
is love and neverless

Notwithstanding the wonders of words, they are inherently limited for studying emotion. The most critical limitation is their temporal relation to experience. When we tell someone how we feel with words, that report is a retrospective reconstruction of an experience. When you report on the delights and frustrations of a day, or your pleasures on a family vacation, or even how a play, art exhibit, or movie moved you, your report is filtered through your current feelings, your intuitive theories of emotional experience, social expectations about what is appropriate to talk about with respect to our inner emotional lives (for example, "how would a mover and shaker express herself here?"), and your personal style (are you prone to repression or dramatic emotional disclosures?). As memories of the emotional experience are dredged up through these filters and then materialize as a set of spoken words, much of the emotional experience remains in the evanescent present of the past, lost. On this, Linda Levine and George Bonanno have found in their research that when people report upon past experiences, be it a disappointing outcome in a presidential election or the death of a loved one, it is their current feelings and how they construe the emotional event that drive their reports of past emotions as much as or more than the original feelings being reported upon.

What was needed was the development of a measure of emotion that approaches the contradictions inherent in a Zen koan. What was needed was an objective, in-the-moment measure that would distill our subjective experiences into unambiguous, quantifiable measures that could be put onto paper and interpreted and debated by scientists. To capture the objective subjective, Ekman and Wallace Friesen devoted seven years, without funding or promise of publication, to developing the Facial Action Coding System (FACS), an anatomically based method for identifying every visible facial muscle movement in the frame-by-frame analysis of facial expression as it occurs in the seamless flow of social interaction. To do so, they boned up on facial anatomy. They trained themselves in the ability to move individual facial muscles (Ekman can roll his eyebrows from one side to the other, like a wave). To document the activity of the more remote muscles in their faces, they stimulated facial muscles layered deep below the surface of the skin with mild electric shock. They then translated how changes in facial appearance— new creases, wrinkles, dimples, bulges—are brought about by different muscle movements, and combinations of muscle movements, into an esoteric language of action units (AUs; see figure below). Ekman and Friesen had given psychological science the first objective mea-sure of specific emotion that could be used in any lab around the world, and in almost any context, as long as the emotional

AU	Description	Facial Muscle	Example Image
1	Inner Brow Raiser	*Frontalis, pars medialis*	
2	Outer Brow Raiser	*Frontalis, pars lateralis*	
4	Brow Lowerer	*Corrugator supercilii, Depressor supercilii*	
5	Upper Lid Raiser	*Levator palpebrae superioris*	
6	Cheek Raiser	*Orbicularis oculi, pars orbitalis*	
7	Lid Tightener	*Orbicularis oculi, pars palpebralis*	
9	Nose Wrinkler	*Levator labii superioris alaquae nasi*	
10	Upper Lip Raiser	*Levator labii superioris*	
11	Nasolabial Deepener	*Zygomaticus minor*	
12	Lip Corner Puller	*Zygomaticus major*	

behavior was videotaped and researchers were manic enough to take the 100 hours to learn FACs and the hour required to reliably code a single minute of behavior.

In the thirty years since this method of measuring facial expression has been developed and distributed to scientists, hundreds of studies have discovered that the muscle configurations that Darwin described for many emotions correspond to the facial expressions people display when feeling the emotion. Ekman's work on facial expression catalyzed a new field—affective science—and led to a more precise understanding of the place of emotion in the brain, the role of emotion in social life, parallels between human and nonhuman emotion, and how we all have different emotional styles. For thirty years, scientists have relied on these methods, and those six emotions, to parse human emotional life. Amid the hundreds of studies, the handbooks, the reviews, the new methodologies and old controversies, one finds empirical support for three deep insights into emotion, the focus of the next chapter. Emotions are signs of our deepest commitments. They are wired into our nervous system. Emotions are intuitive guides to our most important ethical judgments. Our pursuit of the meaningful life requires an engagement with emotion. Our *jen* ratios are revealed in subtle movements of the face.

3

Rational Irrationality

I STILL REMEMBER THE DAY as clear as a bell. Off to the side of the seventh-grade four square game, the love of my life, Lynn Freitas, approached me with hands coyly behind her back. She came unusually close—we were face to face, separated by nine to ten inches—and with a delighted smile framed by her rolling, curly hair asked, "Hey, Dacher, wanna screw?" A surge of thoughts raced through my mind—at last she had recognized my subtle prepubescent allure, at last my longings would make contact with inexplicable happenings on my middle school playground. As I was in the midst of mumbling an earnest and affirmative reply, she held her hand in front of me, palm up, with a screw lying flat on her tender fingers. All I remember was a roar of laughter from the cabal of finger-pointing girls who had suddenly surrounded me to witness this character assassination.

Had I been a FACS-certified seventh grader and known then what I have learned in studying the nonverbal clues of sexual desire (of which Lynn showed not a scintilla), I probably would have been fooled again; it may be in our best interests to be fooled by those we love. Had I trained my ear to discern the fine acoustics involved in playful teasing, I probably would have detected subtle deviations from truthfulness in the artfully elongated vowels of Lynn's enunciation ("Hey, Daaacher, wanna screeew") that would have given away her playful intent. Had I read Nobel Prize-winning economist Thomas Schelling's *The Strategy of Conflict* from 1963, I would have understood why I was fooled, or at least known what to look for when Lynn Freitas made her offer.

Schelling observed that most meaningful exchanges—from the promise of undying love or mutual gain in risky business ventures to the strategic threats of diplomats and negotiators—hinge on solving the commitment problem. The commitment problem has two faces. The first is that we must often put aside self-interested courses of action—offers of affairs, chances to gain at colleagues' expense, opportunities to lie to company stockholders—in the service of our long-term commitments to one another. Long-term relations require that we transcend narrow, in-the-moment, pleasure-seeking self-interest.

The second face of the commitment problem may be more challenging: We must reliably identify who is committed to us, we must find those morally inclined individuals to enter into long-term bonds with, we must know who is (and who is not) likely to be faithful and caring, and disinclined to cheat, lie, and sacrifice us in the service of the pursuit of their self-interest. We must quickly make these decisions to avoid being fooled or exploited by the Lynn Freitases of the world on a regular basis. What helps us solve the commitment problem?

Emotion. The very nature of emotional experience—its seeming absoluteness, heat, and urgency—can readily overwhelm narrow calculations of self-interest, allowing us to honor the commitments integral to long-term bonds: monogamy, fairness, duties and obligations. The potent pangs of guilt help us repair our dearest relations, even at great cost to the self. The single-minded feeling of compassion or awe can motivate us to act on behalf of other individuals or collectives, regardless of costs or benefits to the self.

Just as important is the centrality of others' emotional displays in our attempts to discern others' commitments to us. As important as language is, it is striking how impotent it is in conveying the commitments that define the course of life—the sense that someone will really love us through thick and thin, the sense that a colleague will be a lifelong collaborator, the sense that a politician is devoted to the greater good. Words are easy to manipulate. Not so, emotional displays. Emotional displays provide reliable clues to others' commitments because they are involuntary, costly, and hard to fake (as opposed to words, which Lynn Freitas used to dupe me). Emotional displays have much in common with the peacock's tail or stotting of the red deer (see figure, below): all are metabolically expensive behaviors that are beyond volitional control, and thus less subject to strategic manipulation or deception.

The general claim that Schelling offered: Emotions are involuntary commitment devices that bind us to one another in long-term, mutually beneficial relationships. As Ekman parsed the intricate realm of facial expression, he arrived at a discovery that would provide anatomical support for Schelling's commitment thesis, and that would lead to a rethinking of the centrality of emotion to our most important bonds. Of the forty-three sets of facial muscles, most are easy to move voluntarily. For example, the pictures that follow represent common facial actions, pregnant with signal value, that most anyone can produce at the drop of a hat, at the behest of a friend, to pass the time in the hotel bathroom, or to win a drunken bet.

The peacock's tail and the red deer's stotting display (shown in the presence of predators as a deterrent to the predator's assessment of the likelihood of successfully tracking down the deer) are reliable clues to healthy genes and physical prowess, respectively. These two displays are enormously costly in terms of the energy required to produce, they are impossible to fake, and as a result signal healthy genes to potential mates (the peacock's tail) and physical strength and speed (the deer's stotting).

The three displays above involve simple muscle movements. The display to the far left involves activity of the depressor muscle, which pulls the lip corner down. In the middle one sees activation of the corrugator muscle, which furrows the brow. To the far right one sees movement of the risorious muscle, which pulls the lip corners sideways. All three of these muscle movements can be produced voluntarily. All three expressions have clear signal value. My favorite is the one on the far right, which Ekman proposes is a "referential expression" in that it signals the individual's appreciation that someone else is experiencing an undesired state (that is, the display refers to another's emotion). The display to the far right is often seen when an individual is recognizing another person's experience of fear. But because all three muscle movements are easy to produce voluntarily, they are not critical muscle movements associated with emotion.

A subset of facial muscles, however, are wired differently; they are controlled by different neural pathways originating in the brain. For about 85 to 90 percent of people—actors, sociopaths, politicians, late-night televangelists, and people who take the hundred hours to learn FACS excluded—these muscles are impossible to move voluntarily. If you're feeling bold, want to put some braggart to the test, or are lacking a bit in levity, try yourself or test whether some other poor soul can produce the following muscle actions:

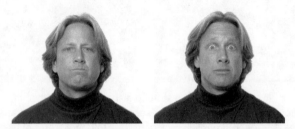

The muscle action on the left signals anger. The muscle combination on the right that pulls the eyebrows up and in, involving the actions of the frontalis and corrugator muscles, signals fear and anxiety.

I've asked dozens of children in summer camps, hundreds of undergraduates in lecture courses, dozens of executives in seminars, most of my indulgent friends, and even, I must confess, my two daughters, to try to produce these muscle actions. After many misfires, contorted faces, shakes of disbelief, and the occasional blush, individuals inevitably fail. What these muscles are, Ekman deduced, are the reliable indicators of emotion. These fleeting movements of muscles in the face are the trustworthy signs of specific emotions, such as anger, fear, desire and love, and, by implication, our social commitments.

Consider sympathy, an emotion central to the stability of the social contract, as Adam Smith, David Hume, and Charles Darwin long ago surmised. Social theorists have homed in on sympathy for some time because it backgrounds the individual's self-interest, and leads to actions that enhance the welfare of others, even at expense to the self. The question is: How do we discern sincere sympathy, or true commitment to others' welfare, from the false promises of demagogues, sociopaths, and hucksters? Robert Frank reasoned, in a synthesis of Schelling's insights and Ekman's methodological labors, that the clues to another's sympathy and commitment to cooperation are found in two simple facial muscle movements (AU1 + AU4, in FACS terminology). Feelings of sympathy, and the commitment to cooperative exchange, are registered in an involuntary facial display that is more trustworthy than its cheap, and readily feigned, copies.

To get a sense of this, compare your reactions to the photos of the two facial displays below. The expression on the left is hard to produce voluntarily. It involves the pulling in and upward of the inner eyebrows, and has been shown in several empirical studies to accompany sympathetic feelings and activation of a region of the nervous system that is associated with caretaking behavior. The facial expression on the right, although quite similar morphologically to the sympathetic display on the left, does not involve activation of these involuntary, reliable facial muscles. It is not a reliable signal of an individual's interest in your welfare (in fact, the eyebrow raise is a signal with many meanings, including interest, skepticism, weakness, and dramatic emphasis when speaking).

The oblique eyebrows of sympathy   Raised eyebrows

Darwin had claimed that our emotional expressions are distilled tokens of more complex social actions—striking out, soothing, eating, embrace, yelling to escape, vomiting, self-protection. Ekman had taken this analysis one step further, showing that of the thousands of possible configurations of facial muscles, a select few are reliable clues to the individual's emotions. This subset of facial expressions, by implication, signals an individual's social commitments, be it likely attack, the inclination to soothe, to be sexually faithful in romantic bonds, or to show concern over social norms and morals.

Emotions feel irrational from the individual's point of view. Emotions can subvert our best attempts at self-control, composure, autonomy, and a narrow self-interested rationality. I'm not at my best at considering the recommendations of a financial advisor, solving crossword puzzles, or sorting out the costs and benefits of my actions when feeling strong emotion.

Long-term relationships, however, require us to put aside utilitarian, cost-benefit analyses of self-interest. Emotions enable us to enact the costly commitments to another's welfare, to respect, to maintaining fair and just relations. Emotions are statements to others that we care, and without these statements long-term relations wither and die. Emotions, Martha Nussbaum argues in *Upheavals of Thought*, are the idiom in which we negotiate our engagements with others. We would live in a lonely, disengaged world were it not for the emotions.

THE SUBLIME BODY

Like many members of his illustrious family, William James was a hypochondriac. It may have been his somatic oversensitivities that led James to publish his radical thesis about emotion in 1884. His thesis turned long-standing intuitions about emotion on their head, and in fact, the role of the head in emotion on its head. Most writers had proposed that our experience of emotion follows from the perception of emotionally evocative events. These experiences, in turn, generate bodily responses rooted in our nervous system. Your experience of embarrassment, for example, follows from your recognition that you've been conducting an important business meeting with toilet paper stuck to your briefcase, and it is this recognition and experience that generate the physiological response—the rush of blood to your cheeks, neck, and forehead that results in the blush.

James's thesis reversed this sequence of bodily response and experience: "My thesis," James proposed, ". . . is that the bodily changes follow directly the perception of the exciting fact, and that our feeling of the same changes as they occur is the emotion." Whereas for Darwin, our repertoire of emotions is wired into our forty-three facial muscles, for James the topography of emotion maps onto our viscera. Every subjective state, from political rage to spiritual rapture to contentment one feels at the sounds of children playing, is registered in its own distinct "bodily reverberation."

Lacking experimental evidence, James turned to thought experiments. One of the most illustrative was the following: What would be left of fear or love or embarrassment, or any emotion, if you took away the physiological sensations such as the heart palpitations, trembling, muscle tensions, warmth or coldness in the skin, sweaty palms, and churning of the stomach? James argued that you would be left with a purely intellectual state. Emotional experience is formed in visceral response.

The bodily system most relevant to James's analysis is the autonomic nervous system, or ANS (see figure below). The most general function of the ANS is to maintain the internal condition of the body to enable adaptive response to ever-changing environmental events. The autonomic nervous system is like the old furnace in a home: It generates energy and distributes it through the body to support our most basic physical activities—digestion, sexual contact, fight or flight behaviors, and just moving the body through space.

The parasympathetic autonomic nervous system incorporates nerves that originate at the top and near the bottom of the spinal cord. The parasympathetic system decreases heart rate and blood pressure, it facilitates blood flow by dilating certain arteries, it increases blood flow to erectile tissue in the penis and clitoris, and it moves digested food through the gastrointestinal tract. The parasympathetic system also constricts the pupil (for feelings of love, look for smaller rather than larger pupils), and it stimulates the secretion of various fluids in the digestive, salivary, and lachrymal glands (for tears). Scientists believe that the parasympathetic branch of the ANS helps the individual relax and restore resources and bodily function. One branch of the parasympathetic ANS originating near the top of the spinal cord—the vagus nerve—is thought to enable caretaking behavior.

The sympathetic autonomic nervous system (ANS) involves over a dozen different neural pathways originating in the spinal cord, and most typically gets the body moving fast. It increases heart rate, blood pressure, and cardiac output. It produces vasoconstriction in most veins and arteries. It shuts down digestive processes. It is associated with contractions in the reproductive organs that are part of orgasm. And it sends fatty acids into the bloodstream, to provide quick energy to the body. The sympathetic ANS helps prepare the body for fight or flight responses.

James's thesis—that each distinct subjective emotion is registered in a different bodily reverberation—is anatomically plausible. The different emotions like disgust, embarrassment, compassion, and awe may originate in different patterns of activation in the heart and lungs, the arteries, and the various organs and glands distributed throughout the body. The first rigorous empirical support for James's claim would arrive 100 years later, in an accidental discovery by Paul Ekman. As Ekman toiled away in his laboratory developing the Facial Action Coding System, he noticed something

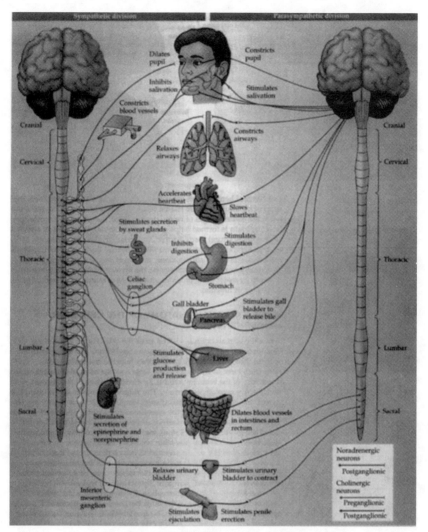

The autonomic nervous system (ANS)

strange. As he moved the different facial muscles to record how they changed the appearance of his face, the different eyebrow positions, nose wrinkles, lip retractions, and the like, these actions actually altered how he felt. When he furrowed his brow, for example, his heart seemed to race and his blood pressure seemed to rise. When he wrinkled his nose, opened his mouth, and stuck out his tongue, his heart seemed to slow, and his stomach felt as if it was turning over. This discovery led him to a striking possibility: that movements of emotional facial muscles stimulate activation in the autonomic nervous system.

What followed was a rather strange and controversial study by Ekman and his colleagues Robert Levenson and Wallace Friesen. It was one of the first to test James's thesis about embodied emotion, using what came to be known as the directed facial action (DFA) task. In this study, participants followed muscle-by-muscle instructions to configure their faces into the six different expressions of the emotions that Ekman had studied in New Guinea. For example, for one expression participants were instructed to:

1. Wrinkle your nose
2. Raise your upper lip
3. Open your mouth and stick out your tongue

Guiding participants to achieve the correct expression required some rather unusual coaching—"no, don't flare your nostrils, instead wrinkle your nose;" "try not to flutter your eyes when bringing your eyebrows up and in"; "as you pull your lips sideways, try not to grit your teeth." Once participants had moved their facial muscles in a fashion that conformed to the required specific emotional expression, they held the expression for ten nerve-racking seconds. During this brief time Levenson recorded several measures of the autonomic nervous system activity associated with the facial expression, which were eventually compared to an appropriate control condition.

The prevailing view was that the ANS is too slow and diffuse to produce emotion-specific patterns of activation. In fact, it was this understanding of the autonomic nervous system that led Schachter and Singer to their vaudevillian study of epinephrine shots and anger and euphoria. The results from this first DFA study refuted this position. These findings would have caused James, a rather shy scholar, living in the shadow of his famous brother, Henry, to blush a little in this empirical confirmation of his controversial thesis. Large increases of heart rate occurred for fear, anger, and sadness, but not disgust. This makes sense given the parasympathetic involvement in digestive processes, which slow the heart down. More subtle was that finger temperature was greater for anger than fear, suggesting that our hot and cold metaphors for anger ("hotheaded") and fear ("cold feet") arise out of bodily sensations.

During moments of anger, blood flows freely to the hands (perhaps to aid in wringing the necks of adversaries), thus increasing the temperature of fingers and toes. During periods of fear, the veins in the arms and legs constrict, leaving much of the blood supply near the chest, which enables flight-related locomotion. It is fair to say, and many critics have, that these distinctions are not the kind of emotion-specific physiological signatures that James envisioned, but these data are certainly a step in that direction.

Levenson and Ekman subsequently packed their physiological equipment up and conducted a similar study with the Minangkabau, a matrilineal Muslim people in West Sumatra, Indonesia. The physiological distinctions between disgust and fear and anger (disgust involves the slowing of heart rate) and between fear and anger (finger temperature is hotter for anger than fear) were once again observed. This result suggests that these linkages between facial expression and autonomic physiology are universal, or at least evident in radically different cultures. And in other research adults aged sixty-five and above show attenuated ANS responses during the DFA, suggesting that, with age, people can more readily move in and out of different emotional states. This parallels studies finding that, as people age, they report experiencing more freedom and control during emotional experiences.

James's unusual thesis inspired other studies of the ANS—of the blush that sears the face, of tears, of goosebumps that ripple down the spine, of the swelling feeling in the chest. These studies reveal that our emotions, even those higher sentiments like sympathy and awe, are embodied in our viscera. As this line of inquiry shifted to the ethical emotions, emotions like embarrassment and compassion, a more radical inference waited on the horizon—that our very capacity for goodness is wired into our body.

THE MORAL GUT

Please read the following passage aloud to people whose moral intuitions matter to you. You might try your family while noshing at the dinner table, old friends reclining in asymmetrical repose around the campfire, or your colleagues sitting around a meeting table, pert and ready. At the conclusion of the passage ask whether they think the person in the passage should be punished or not:

A man goes to the supermarket once a week. On each visit he buys a packaged chicken. He takes it home, draws his curtains, and then has sex with the chicken carcass. He then cooks the chicken and eats it by himself.

What do you think? Lock the person up? Prevent him from coaching little league? Put the individual in handcuffs at the first sight of smoke rising from his barbecue in his backyard? Just ignore this unsettling oddity of his personal life?

When I present this scenario to my students and ask for their punitive judgments, they respond with revulsion. They sit in their seats and recoil reflexively with the full-blown Darwinian, Jamesian emotion of moral disgust written across their faces in raised upper lips and flared nostrils and felt in the visceral turning of the stomach and the slowing of heart rate. Then, like good students of western European culture, they recall their civics lessons about individual rights, freedoms, and privacy. They eventually decide, their viscera notwithstanding, that the individual should not be punished; he should have the right to practice such a culinary (or sexual) act in the privacy of his own home, as long as he has curtains closed and refrains from writing cookbooks or having friends over for dinner.

People's responses to this kind of thought experiment have led Jonathan Haidt to a new view of moral judgment, and one that prioritizes the moral gut. Haidt argues that our moral judgments of right and wrong, virtue, harm, and fairness, are the products of two kinds of processes. The first may seem fairly intuitive to you—it has occupied the thinking of those who have theorized about moral judgment for 2,000 years—and that is complex, deliberative reason. When we judge whether an action is right or wrong, we engage in many complex reasoning processes, we consider society-wide consequences, cost-benefit analyses, motives and intentions, and abstract principles like rights, freedoms, and duties. Psychological science has privileged these higher-order reasoning processes in accounts of moral judgment. This is no better typified than by the well-known theory of moral development of Harvard psychologist Lawrence Kohlberg. Beginning with his dissertation, Kohlberg argued that the highest forms of moral judgment require abstract considerations of rights, equality, and harm—achieved, in his research, by only 2 to 3 percent of individuals he studied around the world (most typically highly educated, upper-class males like himself!).

The second, more democratic element of moral judgment, almost completely ignored in psychological science, is the gut. Emotions provide rapid intuitions about fairness, harm, virtue, kindness, and purity. When you first reacted to the sex-with-chicken example, part of your response was most likely a rapid, ancient feeling of revulsion and disgust at the image of such a species-mixing, impure sexual practice. In one study, my first mentor, Phoebe Ellsworth, and I had individuals move their facial muscles, much as Ekman and colleagues did with the DFA, into the facial expression of anger or sadness. As participants held the expression, they made quick judgments about who was to blame for problems they might experience in the future in their romantic, work, and financial lives—other people or impersonal, situational factors. Those participants who made these judgments with an angry expression on their face blamed other people for the injustices. Those with faces configured into a sad expression attributed the same problems to fate and impersonal factors. Our moral judgments of blame are guided by sensations arising in the viscera and facial musculature.

Haidt reasons that thousands of generations of human social evolution have honed moral intuitions in the form of embodied emotions like compassion, gratitude, embarrassment, and awe. Emotions are powerful moral guides. They are upheavals that propel us to protect the foundations of moral communities—concerns over fairness, obligations, virtue, kindness, and reciprocity. Our capacity for virtue and concern over right and wrong are wired into our bodies.

If you are not convinced, consider the following neuroimaging study of Joshua Greene and colleagues, which suggests that the emotional and reasoning elements of moral judgment activate different regions of the brain. Participants judged different moral and nonmoral dilemmas in terms of whether they considered the action to be appropriate or not. Some moral dilemmas were impersonal and relatively unemotional. For example, in the "trolley dilemma" the participant imagines a runaway trolley headed for five people who will be killed if it proceeds on its course. The only way to save them is to hit a switch that will turn the trolley onto an alternate set of tracks, where it will kill one person instead of five. When asked to indicate whether it's appropriate or not to hit that switch and save five lives, participants answer yes with little hesitation.

As an illustration of the emotionally evocative scenarios, consider the "footbridge dilemma." Again five people's lives are threatened by a runaway trolley. In this case the participant imagines standing next to a very heavy stranger on a footbridge over the trolley tracks. If the participant pushes the rotund stranger off the bridge with his own hands and onto the tracks, the stranger dies, but the train veers off its course, thus saving five lives (the participant's own weight, it is explained, is insufficient to send the trolley off the track). Is it appropriate to push the stranger off the footbridge?

While participants responded to several dilemmas of this sort, functional magnetic resonance imaging techniques ascertained which parts of the participant's brain were active (see figure below). The personal moral dilemmas activated regions of the brain that previous research had found to be involved in emotion. The impersonal moral dilemmas and the nonmoral dilemmas activated brain regions associated with working memory, regions centrally involved in more deliberative reasoning.

When the Dalai Lama visited the gas chambers of Auschwitz, and reflected, stunned, upon this most horrific of human atrocities, he offered the following: "Events such as those which occurred at Auschwitz are violent reminders of what can happen when individuals—and by extension, whole societies—lose touch with basic human feeling." His claim is that the direction of human culture—toward cooperation or genocide—rests upon being guided by basic moral feelings. Confucius was on the same page: "the ability to take one's own feelings as a guide—that is the sort of thing that lies in the direction of *jen*." Martha Nussbaum, bucking the trends of

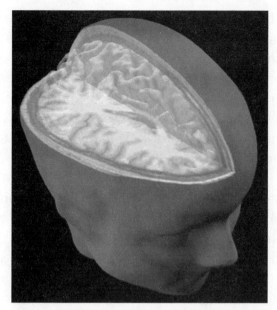

Greene's more emotionally evocative moral dilemmas, such as the footbridge dilemma, stimulated
activation in the two midline regions of the brain, the
posterior cingulate, as well as the superior temporal
sulcus, located to the back of the brain. The more
impersonal moral dilemmas stimulated activation in
the dorsal, anterior region of the brain (to the front
and right).

moral philosophy, concurs by arguing that emotions at their core
contain judgments of value, about fairness, harm, rights, purity,
reciprocity—all of the core ideas of moral and ethical living. Emotions are guides to moral reasoning, to ethical action in the fast,
face-to-face exchanges of our social life. Reason and passion are collaborators in the meaningful life.

ENEMIES NO MORE

We often resort to thought experiments to discern the place of emotion in social life. Natural-state thought experiments plumb our
intuition about what humans were like prior to culture, civilization,
or guns, germs, and steel. Ideal-mind thought experiments—used
in meditation and in philosophical exercises like the moral philoso-

pher John Rawls's veil of ignorance—ask us to envision the mind operating in ideal conditions, independent of the press of our own desires or the web of social relations we find ourselves in.

Emotions have not fared well in these thought experiments. Philosophers have most consistently argued that emotions should be extirpated from social life. This train of thought finds its clearest expression in the third century BC, with the Epicureans and Stoics; it extends to St. Augustine, St. Paul, and the Puritans, and on to many contemporary accounts of ethical living (for example, Ayn Rand). In the words of the influential American psychologist B. F. Skinner: "We all know that emotions are useless and bad for our peace of mind."

If this brief philosophical history seems a bit arcane, consider the metaphors that we routinely use in the English language to explain our emotions to others (see table below), revealed by linguists Zoltán Kövecses and George Lakoff. We conceive of emotions as opponents (and not allies). Emotions are illnesses (and not sources of health). Emotions are forms of insanity (and not moments of understanding). We wrestle with, become ill from, and are driven mad by love, sadness, anger, guilt, shame, and even seemingly more beneficial states like amusement. The opposite locks up the Western mind: Imagine referring to anger, love, or gratitude as a friend, a form of health, or a kind of insight or clarity. We assume that emotions are lower, less sophisticated, more primitive ways of perceiving the world, especially when juxtaposed with loftier forms of reason.

METAPHORS OF EMOTIONS

Emotions = Opponents	*I'm wrestling with my grief* *My enthusiasm got the best of me* *I couldn't hold back my laughter*
Emotions = Disease	*I'm sick with love*
Emotions = Insanity	*He's mad with rage*

As Paul Ekman began to publish his work on the Foré, his papers set in motion a scientific revolution that required a radical revision of time-honored assumptions about human nature. This science began to uncover how emotions are wired into our facial anatomy, our vocalizations, our autonomic responses, and our brains. We learned that emotions support the commitments that make up the social contract with friends, romantic partners, siblings, and offspring. Emotions are not to be mastered by orderly reason; they are rational, principled judgments in their own right. Emotions do not subvert ethical living; they are guides to moral action, and they tell us what matters. Emotions like compassion, embarrassment, gratitude, and awe are the substance of high *jen* ratios and the meaningful life.

Deeper insights into the origins of the emotions—the very question that spurred Darwin to write *Expression* and that led Ekman to New Guinea—would come from new insights into the nature of human evolution. This new evolutionary literature, the topic of our next chapter, would reveal that those hominid predecessors guided by emotions such as compassion, embarrassment, and awe fared better in the tasks of survival, reproduction, and raising offspring to the age of viability. Evolution, it would seem, smiles upon those with higher *jen* ratios.

4

Survival of the Kindest

I N NOVEMBER 1943, S. L. A. "Slam" Marshall, a U.S. Army lieutenant colonel, arrived with American troops on the beaches of Makin Island to fight the Japanese. After four days of bloody, chaotic combat, the Americans secured the island. In the ensuing calm, Marshall was asked to interview several soldiers to clarify some specifics of the four-day battle, with medals, heroic claims, and rights to wartime stories at stake. Marshall subsequently interviewed hundreds of soldiers who fought in Europe and the Pacific during World War II, often immediately after engagement. In 1947, he published the results from these interviews in *Men Against Fire: The Problem of Battle Command*.

His interviews yielded an astonishing finding: Only 15 percent of World War II riflemen had fired at the enemy during combat. Often soldiers refused to fire at the enemy with superior officers barking commands nearby and bullets zipping past their heads. In the wake of this revelatory finding, the army radically changed how it prepared soldiers to kill. Infantry training exercises played down the notion that shooting kills humans. Soldiers were taught to shoot at nonhuman targets—trees, hills, bushes, cars, hovels, huts. The effects were dramatic. According to army estimates, 90 percent of soldiers in the Vietnam War fired at their enemies.

If Charles Darwin and his close intellectual peers—Thomas

Henry Huxley and Alfred Russel Wallace—were to discuss this finding with Charlie Rose or on C-SPAN—that in the heat of battle soldiers most typically refused to harm fellow human beings in spite of their self-preservation being on the line—they would reach contrasting conclusions. For Alfred Russel Wallace, a codiscoverer of the theory of evolution by natural selection, this concern for the welfare of others would be taken as evidence of how God has shaped human beings' more benevolent tendencies. Wallace argued that while the body was shaped by natural selection, our mental faculties, and most notably our capacity for good, were created by "an unseen universe of the Spirit" (p. 354). It was some kind of spiritual force that kept soldiers from pulling the trigger to end the lives of enemies.

T. H. Huxley, progeny of one of England's well-known intellectual families, was evolutionary theory's fiercest early advocate and public spokesman. In Oxford and Cambridge circles he was nicknamed Darwin's bulldog. He would have readily attributed Marshall's findings to the constructive forces of culture. In Huxley's view, human nature is aggressive and competitive, forged by evolution in a violent, selfish struggle for existence. Altruistic actions oriented toward benefiting the welfare of others—soldiers refusing to harm, daily civilities of public life, kindness toward strangers—must be cultivated by education and training. Cultural forces arise to counteract the base instincts that evolution has produced at the core of human nature.

Darwin would have reached yet a different conclusion, parting ways with his two colleagues. Had he been able to do so, he might have placed Marshall's empirical gem in his first book on humans, *Descent of Man*, published twelve years after the *On the Origin of Species*. In *Descent*, Darwin argued that the social instincts—instincts toward sympathy, play, belonging in groups, caring for offspring, reciprocating acts of generosity, and worrying about the regard of others—are part of human nature. In Darwin's typically modest but provocative prose:

The following proposition seems to me in a high degree probable—namely, that any animal whatever, endowed with well-marked social instincts, the parental and filial affections being here included, would inevitably acquire a moral sense or conscience, as soon as its intellectual powers had become as well, or nearly as well developed, as in man. For, firstly, the social instincts lead an animal to take pleasure in the society of its fellows, to feel a certain amount of sympathy with them, and to perform various services for them. The services may be of a definite and evidently instinctive nature; or there may be only a wish and readiness, as with most of the higher social animals, to aid their fellows in certain general ways. But these feelings and services are by no means extended to all the individuals of the same species, only to those of the same association. . . . after the power of language had been acquired, and the wishes of the community could be expressed, the common opinion how each member ought to act for the public good, would naturally become in a paramount degree the guide to action. But it should be borne in mind that however great weight we may attribute to public opinion, our regard for the approbation and disapprobation of our fellows depends on sympathy, which, as we shall see, forms an essential part of the social instinct, and is indeed its foundation-stone. Lastly, habit in the individual would ultimately play a very important part in guiding the conduct of each member; for the social instinct, together with sympathy, is, like any other instinct, greatly strengthened by habit, and so consequently would be obedience to the wishes and judgment of the community.

Our moral capacities, Darwin reasoned, are rooted in sympathy. These capacities are constrained by association or familial relatedness (anticipating what would come to be called, nearly 100 years later, kin selection theory). They are strengthened by habit and social practice. Later, in explaining acts of altruism, Darwin makes an even stronger claim:

Such actions as the above appear to be the simple result of the greater strength of the social or maternal instincts than that of any other instinct or motive; for they are performed too instantaneously for reflection, or for pleasure or pain to be felt at the time; though, if prevented by any cause, distress or even misery might be felt. In a timid man, on the other hand, the instinct of self-preservation, might be so strong, that he would be unable to force himself to run any such risk perhaps not even for his own child.

Our evolved tendencies toward goodness, Darwin proposed, are performed with the automatic, well-honed speed of other reflexes—the flinch of the body at a loud, unexpected sound, the grasping reflex of the young infant. They are stronger than those toward self-preservation, the default orientation of timid men. Darwin's early formulations of the social instincts of humans were clearly tilted toward a positive *jen* ratio, where the good is stronger than the bad.

CRO-MAGNON FIELD NOTES

There are many books I would love to read but, alas, never will: the autobiography of Jesus; a stream-of-consciousness narrative of Virginia Woolf's last thoughts as she plunged into the River Ouse, weighed down by heavy rocks tucked into her coat pockets. As alluring as those certain best sellers would be, at the top of my list would be the field notes of a Cro-Magnon anthropologist, who would have had the wherewithal to travel through Africa, Europe, and Asia to characterize the social life of our most immediate hominid predecessors some 30,000 to 50,000 years ago.

A detailed portrayal of the day in the life of our hominid predecessors would shed light on our environment of evolutionary adaptedness (EEA). The EEA is an abstract description of the social and physical environment in which the human species evolved. It is within this environment that certain genetically based traits—for example, to avoid foods with foul odors that signal decay, to respond with charm and sexual readiness when a female is ovulating—led to greater success in the games of survival and reproduction, and became encoded into the human genome, while others led to increased probabilities of fatality and cold shoulders from potential mates, and quickly to the scrap heap of evolution.

These Cro-Magnon field notes would flesh out Darwin's early evolutionary analysis of our moral capacities. A clear picture of early hominid social life would tell us of the recurring social contexts that reduce the chances of genes making it to the next generation—the perils of escalated aggression between males, the prevalence of infidelity and strategic cuckoldry, the reduced likelihood of offspring surviving if fathers are not engaged. We would also read of the social tendencies that increase the chances of gene replication—the sharing of food or caring for offspring, social strategies that allow females and males to rise in social hierarchies, thereby gaining preferential access to resources and mates. Knowing these social facets of the EEA would then lay a platform for understanding the deeper origins of where the blush of embarrassment comes from, why we can communicate pro-social emotions like gratitude or compassion by one-second touches to a stranger's arm, how devoted love is represented in the flow of certain neuropeptides in the bloodstream.

Absent these Cro-Magnon field notes, we can turn to several kinds of evidence, and a Darwinian capacity for going beyond the information given, to envisage our EEA. We can turn to studies of our closest primate relatives, chimpanzees and bonobos in particular—with whom the human species shared a common ancestor some seven to eight million years ago. Here similarities in social existence—caregiving or hierarchical organization, for example—tell us about basic primate social tendencies and the organization of the pro-social branches of the nervous system. Differences—for example in pair bonding patterns—uncover likely sources of the specifics of human design, and new dimensions to our emotional life.

We can turn to the scanty archaeological record of human ancestry. Here exciting debates are clarifying the meaning of piles of animal bones near ancient hearths, shifts in skeletal structure in the predecessors of *Homo sapiens*, and the first attempts at visual art and music. From these debates we are learning some basic facts about our hominid predecessors.

Finally, we can rely on the detailed observations of contemporary hunter-gatherer societies in remote places in the Amazon, Africa, and New Guinea. These rich descriptions of hunter-gatherer social life—studies of the !Kung San of southern Africa, for example—provide hints into what day-to-day life might have been like for our hunter-gatherer predecessors tens of thousands of years ago.

If we had those Cro-Magnon field notes, we would read that our hominid predecessors spent most of their minutes alive in the presence of other group members, living in close proximity in thirty- to seventy-five-person groups. Division of labor was pronounced: Females served as the primary gatherers of food and caretakers of infants during the extended period of immaturity, traveling less than males, who would have devoted much of their time to the tasks of hunting—flaking stones for weapons, carving spears, tracking game, sharing information about migration patterns and moments of prey vulnerability. Our Cro-Magnon writer, though, would have to have taken note of the relative similarity in size of females and males (the average difference in size between modern human males and females is about 15 percent; in the hominid species that preceded our immediate predecessor, males were about 50 percent bigger than females) and the competition between males for access to mates that would have produced this leveling off of size differences between the sexes.

Several chapters would reveal the darker side of our hominid predecessors, and the origins of the disturbing tendencies of contemporary humans. Here the Cro-Magnon anthropologist would have ample data to write about the regularity of male-on-male violence. There would be extensive observations about warlike behavior, and raids on other groups that might give rise to murder and rape. The regularity of strategic infanticide would emerge as a theme.

At the same time, our Cro-Magnon anthropologist would write specific chapters about social dimensions of the lives of our hominid predecessors that would illuminate the origins of emotions like embarrassment, compassion, love, and awe, and our early capacity for *jen*.

Stages of human evolution

TAKE CARE OR DIE

The first chapter of the Cro-Magnon field notes would be devoted to the prevalence of caregiving, a hallmark feature of higher primates. As Frans de Waal has observed, chimpanzees and bonobos often become wildly distressed when witnessing harm to other group members. Chimps and bonobos routinely protect conspecifics born blind. They shift their play, resource allocations, and physical navigation of the environment when interacting with fellow primates crippled by physical abnormalities. They, like us, are attuned to harm and vulnerability, and tailor their actions accordingly.

Caregiving is all the more pressing an adaptation in hominids, thanks to shifts in the composition of our predecessors' social groups. Studies of our predecessors' bones reveal that for the first time in primate history, our predecessors were often living into old age, up to the age of sixty. These first older primates, wise with information about food sources, how to care for offspring, and climate patterns, likely required care from younger members of the group.

Even a more pervasive and pressing fount of caregiving was the radical dependence of our hominid predecessors' offspring. Our hominid predecessors evolved bigger brains: *Homo erectus* had brains about 1,000 cc, which is 50 percent bigger than those of their immediate predecessors, *Homo habilis*. The females evolved narrower pelvises, which emerged to support upright walking as our predecessors descended from arboreal life to become bipedal omnivores on the African savannah. As a result, early hominids were born premature, to squeeze through the narrower pelvis region. They entered the world with big brains but few physical survival skills. They had a longer period of dependency than those of their primate predecessors, and required more care, so much so that our hominid social organization had to shift radically, as did our nervous systems.

In his review of hunter-gatherer social life in The *Tangled Wing*, Melvin Konner notes the pervasiveness of intensive infant care. This care was typically provided by mothers, but also by engaging fathers, younger female relatives (aunts, sisters), and younger children. Such care might seem indulgent to our modern, Benjamin Spock-trained sensitivities, but in hunter-gatherer culture, it was a given. Thus, Konner observes that the !Kung infant is

> carried in a sling at the mother's side, held vertically in continuous skin-to-skin contact. Reflexes such as crawling movements in the legs, the use of the arms to move and free the head, and grasping responses in the hands allow the infant to adjust to the mother's movements and avoid smothering in her skin and clothing. These movements also signal the infant's changes of state, teaching the mother to anticipate its waking, hunger, or defecation. The hip position lets infants see the mother's social world, the objects hung around the neck, any work in her hands, and the breast. Mutual gaze with the mother is easy, and when she is standing the infant's face is just at the eye level of keenly interested ten- to twelve-year olds, who frequently initiate brief, intense, face-to-face interactions. When not in the sling, infants are passed from hand to hand around a fire for similar interactions with adults and children. They are kissed on their faces, bellies, and genitals, sung to, bounced, entertained, encouraged, and addressed at length in conversational tones before they can understand words.

Infant bonobos (*left*) and chimpanzees (*right*) evoke similar oohs and ahhs as our own infants, thanks to their heartwarming cuteness. But they achieve self-sufficiency much earlier than human infants, learning much more quickly to eat, find shelter, and navigate the physical environment on their own.

The mother indulges the infant's dependency completely in the first year and the second year resists it only slightly. Nursing is continual, four times an hour throughout the day on average, triggered by any slightly fretful signs. Close contact for the first two years allows for a much more fine-grained responsiveness by the mother than can be attained in a culture where mother and infant are often apart.

Caregiving is a way of life in humans, and has been wired into our nervous system in the forms of emotions, such as sympathy and filial love.

FACE TO FACE

A second feature of early humans' social EEA is that it was almost continually face-to-face. Don't be misled by the hours you spend alone, commuting, on the Internet, on your cell phone, or fingering your BlackBerry while eating in your car. The amount of time we spend alone is a radical aberration for our species (and a source of many contemporary social and physical ills). Early humans required one another to accomplish the basic tasks of survival and reproduction. They did so in highly coordinated, face-to-face interactions. Cooperative child rearing, where relatives and friends traded off duties, was central to quotidian life, as hinted at in Konner's quotation above.

Studies of archaeological sites reveal consistent evidence of cooperative hunting for meat—a critical part of early hominid diet. Relative to many of the animals early humans hunted—bison, elephants, rhinoceroses—our predecessors were weak and slow of foot, and lacking in the fangs, claws, speed, and strength seen in other predators. Early hominid strength was found in coordination and cooperation. For example, at Mauran, in the French Pyrénées, a massive accumulation of bison bones near a river, thought to be 50,000 years old, suggests that teams of Neanderthals banded together to force herds of bison off cliff edges, to fall to their deaths.

The continual coordination required of early human social life coevolved with morphological changes that gave rise to our remarkable capacity to communicate, which is unlike that of any other species in terms of precision, flexibility, sensitivity, and band width. Unlike our primate relatives, the human face has relatively little obscuring hair (which most likely was lost in the hot African savannah, for purposes of cooling), making it a beacon of social messages. And our facial anatomy includes more facial muscles than those of our primate relatives, in particular around the eyes, allowing for a much richer vocabulary of expressive behavior originating in the face.

The evolving capacity to communicate is even more pronounced in the human voice. With emerging bipedalism in our hominid predecessors, the human vocal apparatus evolved dramatically. Compared to our primate predecessors, the human vocal tract is elongated. As a result, the tongue has greater range of movement at the back near the larynx, allowing for the capacity to produce a remarkable variety of sounds. Some of the great apes, for example, have an extremely limited repertoire of vocalizations, which reduces to a few grunts. Humans, in contrast, can exhort, punish, threaten, tease, comfort, soothe, flirt, and seduce with the voice.

Our evolving capabilities to communicate co-evolved with our broader capacity for culture, our tendency to produce artifacts, to imitate, to represent and spread information across time and space with language. As charming as chimps and bonobos are, careful studies of their social existence find little evidence of anything remotely resembling culture—a point many have recently made in suggesting that the human capacity for imitation, symbolic language, memory, and coordination is radically different from that of our primate relatives. In humans our basic emotional tendencies can quickly spread to others, through mimicry, imitation, and communication. The spread of emotions like compassion, love, and awe becomes the basis for social ritual and ethical guidelines, and binds individuals into cooperative groups.

CRO-MAGNON CEOS

Our Cro-Magnon anthropologist would readily discern a third feature of early human social life—that it is hierarchical. Every moment of early hominid social life, from who sleeps with whom to who eats what to who touches whom, was stratified. In contemporary humans, individuals fall into social hierarchies with remarkable ease. In research with my colleague Cameron Anderson, we have found that hallmates, within one week of having moved into college dormitories, are nearly unanimous in singling out those whom they report to be of high status, having respect, prominence and influence in their emergent groups. They likewise readily agree in their judgments of who occupies the lower rungs of the totem pole. Differences in status quickly emerge in younger children (down to two years old, where status hierarchies have been observed on the seemingly egalitarian circle rugs of preschools). And don't be fooled by gender-based assumptions: The concern for status is not just a male thing. Female adults attain comparable levels of status with just as much alacrity and effect. This is echoed in recent studies by Frans de Waal and others, who have documented clear hierarchies in female chimpanzee life. Primate social life is hierarchical, in large part because hierarchies enable group members to decide how to allocate resources with speed and minimal conflict.

Yet the hierarchical social organization of higher primates and early humans differs dramatically from that of other species. In higher primates and humans, lower-status individuals can readily form alliances, most typically dyadic coalitions, which potentially negate many advantages that higher-status individuals might enjoy in physical size or power. In addition, humans developed several forms of social communication—for example, gossip—by which low-status individuals can comment upon and determine the status of other group members. The emergence of coalitions and alliances in group life, and the capacity for low-status individuals to comment on the reputations of those in power, placed new demands upon high-power individuals. Their power would come to rest increasingly upon the ability to engage socially and advance the interests of the group.

Frans de Waal has found in his groundbreaking studies of primate politics that with the rise of the capacity of lower-status individuals to form coalitions, "alpha" males and females must rely on social intelligence to acquire and retain their privileged positions. Pure intimidation displays—chest pounding, random fang-bearing charges, throwing stones, and din making—are and were still stock-in-trade for alpha chimps and bonobos and our human predecessors, but new skills were required. Higher-status primates spend a great deal of their day smoothing over the rough edges of their group's social existence. They are the ones who are likely to mediate conflicts, for example by bringing adversaries into physical contact with one another and encouraging grooming activities that reduce conflict. They are the ones who make sure that more equitable allocations of resources occur.

My own research with humans paints a similar picture. We have studied who quickly rises in male and female hierarchies in groups of children and young adults. We find that it is not the domineering, muscle-flexing, fear-inspiring, backstabbing types who gain elevated status in the eyes of their peers (apologies to Machiavelli). Instead, it is the socially intelligent individuals who advance the interests of other group members (in the service of their own self-interest) who rise in social hierarchies. Power goes to those who are socially engaged. It is the young adults and children who brim with social energy, who bring people together, who can tell a good joke or tease in ways that playfully identify inappropriate actions, or soothe another in distress, who end up at the top. The literature on socially rejected children finds that bullies, who resort to aggression, throwing their weight around, and raw forms of intimidation and dominance, in point of fact, are outcasts and low in the social hierarchy. Power and status are inevitable facets of hominid social life but are founded on social intelligence more than Social Darwinism.

THE PERPETUAL CONFLICT OF BEING

Lest you suspect that our Cro-Magnon anthropologist suffers from a Pollyanaish view of her own kind—a universal bias of most human groups—it is wise to consider her fourth generalization. Here she would observe that almost every waking second of early hominid social life is pervaded by continual and often painful conflict.

There would be discussion of obvious within-sex conflicts, for example, over mates and resources. Early hominid social organization increasingly came to revolve around the competition between males for access to females. The same applies to females, who, as Darwin long ago surmised, adorn and beautify themselves in an arms race of beauty to attract resource-rich mates.

This logic of competing interests extends to parent-offspring relations, as Sarah Blaffer Hrdy brilliantly shows in *Mother Nature*. Offspring make competing demands upon parents. As a result, parents are required to make strategic and often disarmingly utilitarian judgments about which offspring to devote resources to, and, in extreme circumstancs such as famine (or in today's political climate, civil war), which to abandon.

This parent-offspring conflict even extends to mother-fetus relations, as Harvard evolutionary biologist David Haig has demonstrated at the genetic and physiological levels. Many of the pathologies of human pregnancy—hypertension and diabetes, for example—have been newly understood from the perspective of the fetus's making self-serving demands upon the mother's supply of nutrients, at considerable cost to the mother.

Siblings are not safe from perpetual, and occasionally mortal, conflict. I remember late one night preparing for a lecture on family dynamics and moral development, having just put my daughters, Natalie, then 4, and Serafina, then 2, to bed. As they peacefully slept in their splayed-out positions, as if dropped out of space onto their beds next to one another, I encountered a fact that left me in shoulder-slumping laughter and tears. In an observational study of American families, four- and two-year-old siblings were observed to engage in conflicts—eye poking, name calling, hair pulling, toy grabbing, arm biting, cheek scratching—every eleven minutes of waking existence.

This kind of sibling conflict, Frank Sulloway reveals in *Born to Rebel*, is expected, based on evolutionary theory. Siblings share, on average, 50 percent of their genes, and compete over numerous resources, from the protection and affection of parents to food to mates—particularly when resources are scarce. Sibling conflict is frequent, widespread, and, on occasion, deadly. Sibling sand sharks devour one another prior to birth in the oviducts of the mother, until one well-fed shark emerges. Once a blue-footed boobie drops below 80 percent of its body weight, its siblings exclude it from the nest, and at times will peck it to death. Infant hyenas are born with large canine teeth, which they often turn to deadly effect upon their newly born siblings.

Conflict is synonymous with human social life. Yet early hominid conflict differed from that of many other species: It was met with evolved capacities to reconcile. This essential insight can be traced back to the observations of Jane Goodall and Frans de Waal, who documented how our primate relatives reconcile after aggressive encounters. Prior to Goodall and de Waal's work, the prevailing wisdom, developed by ethologist Konrad Lorenz, was that following an aggressive encounter, aggressors moved away from each other as far as possible. This view might make sense for solitary species, like the golden hamster, who flee upon attack, or territorial species, like many birds, who rely on birdsong to create invisible but audible property lines to avoid deadly conflicts.

For many mammals, though, these options—fleeing the group or solitary territorial arrangements—do not make evolutionary sense. Our hominid predecessors were dependent upon one another to defend against predators, hunt, reproduce, and ensure that offspring reached the age of viability and reproduction. Individuals who were better able to negotiate conflicts almost certainly fared better in the tasks of survival and gene replication. Recent studies have found that wolves who have been kicked out of their group for excessive aggression and an inability to play are less likely to reproduce and more likely to die. Many physiological difficulties associated with human isolation—namely, increased stress, weaker responses to disease, and even shorter lives—suggest that our survival depends on healthy, stable bonds with others. Conflict is costly and painful but better than the alternative—a solitary existence of fending for oneself. Out of the perpetual conflict that runs through human social life emerged a rich array of capacities that short-circuit or defuse conflict—appeasement displays, forgiveness, play, teasing, and laughter.

FRAGILE MONOGAMY AND THE NEW DAD

Finally, our Cro-Magnon anthropologist would have to devote a surprisingly chaste chapter to the bawdy politics of our primate predecessors. Their sexual organization differs from that of our closest primate relatives, and makes us resemble the local Towhee or warbler flitting about rather than baboons or chimpanzees. We are relative prudes compared to these primate relatives. Once a female chimpanzee is sexually mature at age fifteen, she advertises her sexual receptiveness by a large pink patch of sexual skin, and for a ten-day period during a thirty-six-day menstrual cycle, she copulates several dozen times a day, with all or most of the adult males in her social group. Aggression and jockeying for access to female chimpanzees during these periods become all-consuming for male chimpanzees. Females raise offspring largely on their own; males contribute to the community but not to individual offspring, and males don't know which offspring they have fathered.

Then there are the bonobos, now recognized as a separate species from chimpanzees, and widely envied by humans yearning for the next sexual revolution. Bonobo females are sexually active for about five years before they become fertile, and copulate freely with many of the adult males in their immediate social group. Female and male homosexual relations are common. Younger males often engage in sexual activity with older females in what looks like sexual initiation play. Sexual contact among the bonobos is the basis of friendships, conflict reduction, and play.

The monogamous tendencies documented by our Cro-Magnon anthropologist, by contrast, are unusual for higher primates. They have never been observed, except in hominids, in species where the sexes mix in large groups without territorial boundaries. This sexual organization had several important implications. Females evolved to become sexually active throughout their menstrual cycles. Males and females could maintain exclusive sexual interest in each other. For example, in a survey of world cultures, monogamy was recognized as official policy in only 16 percent of 853 societies sampled, but sexual monogamy was the most common sexual pattern. Males evolved to know who their offspring were and to provide resources and care to them.

Our Cro-Magnon anthropologist, then, would conclude that the social environment of the EEA would be defined by an acute tendency to care, by highly coordinated, face-to-face social exchanges, by the need to reconcile and the flattening of social hierarchies, by perpetually negotiated conflicts of interests, and by the emergence of the tendency toward sexual monogamy. It is these properties of our early social existence that gave rise to the moral emotions, of interest to Darwin but long ignored by the science of emotion that he inspired. Compassion, embarrassment, awe, love, and gratitude emerged in the recurring social interactions of early hominid social life: the attending to vulnerable offspring, the playful exchanges between kith and kin, the status moves and negotiations, the courtships and flirtations between current and potential sexual partners. These emotions were wired into the body and our social life through processes of natural and sexual selection. They evolved into the language of human social life, the species-characteristic patterns of parent-offspring relations, relations between mates and allies, dominant and subordinate members of hierarchies, and in mating relationships. These emotions became our ethical guides to help us fold into stable, cooperative communities. They operated according to three general principles, revealed in a tournament that pitted the brightest mathematicians and computer hacks against one another in an attempt to discover what strategies prevail in the survival of the fittest.

THE WISDOM OF TIT-FOR-TAT AND THE GREAT SHIFT

In *The Evolution of Cooperation* Robert Axelrod asks the following question: How might cooperation emerge in competitive environments governed by the ruthless pursuit of self-interest? How might compassion, awe, love, and gratitude, powerfully oriented toward enhancing the welfare of others, take hold within social communities governed by the pursuit of self-interest, in such a fashion that they would become favored by natural selection and encoded into our genes and nervous system?

Axelrod himself was taken aback by striking acts of cooperation that confound assumptions about self-preservation and self-interest. In the trenches of World War I, for example, British and French soldiers were separated from their enemies, the Germans, by a few hundred yards of burned-out, treeless, muddy no-man's-land. Brutal assaults by one side were typically met with equally fierce, lethal attacks by the other. And, yet, in these nightmarish patches of annihilation, cooperation emerged. The two sides flew certain special flags, signaling nonconfrontation. They made verbal agreements not to shoot at each other. They evolved patterns of firing their weapons in purely symbolic, harmless ways, to signal nonlethal intent. All of these cooperative strategies allowed the soldiers to eat meals peacefully and to enjoy long periods of nonengagement. On special occasions, the warring sides even fraternized with one another. In fact, cooperation became so pervasive that commanding generals had to intervene, demanding a return to deadly combat.

From historical anecdote Axelrod turned to the prisoner's dilemma game (see table below) to answer his question about the evolution of cooperation. He conducted a tournament in which players—cold war strategists, psychologists, prize-winning mathematicians, computer specialists, and other aficionados of the game—were invited to submit computer programs that specified what choice to make on a certain round of the prisoner's dilemma game, given what had happened in previous rounds. In Axelrod's first tournament, fourteen different strategies were submitted. Each was subsequently pitted against the others for 200 rounds. Here the game really mirrors human social life. Individuals with different strategic approaches went toe-to-toe with one another, much as bullies and altruists do on the grammar-school playground, Machiavellians and kindhearted colleagues do at work, hawks and doves do in foreign policy debate, and presumably our hominid predecessors—genetically prone, through random mutation, to cooperate or compete—did. Who prevailed?

Confounding assumptions about self-preservation and self-interest, British and French soldiers fraternized with the enemies, the Germans, during the bitter trench warfare of World War I.

THE PRISONER'S DILEMMA GAME (PDG)

		PARTNER'S ACTION	
		COOPERATE	COMPETE
YOUR ACTION	COOPERATE	5,5	0,8
	COMPETE	8,0	2,2

In the PDG, participants are required to make a simple choice: to cooperate or compete with one another. If both participants cooperate, they do well (in our example, they each receive $5). If one competes while the other cooperates, the competitor thrives at the expense of the cooperator (in our example receiving $8 to the cooperator's zilch). If both compete, they each get $2. From the perspective of maximizing self-interest, the rational thing to do is to compete. The rub, though, is that, as in arms races, the use of shared resources, intimate life, and business partnerships, the mutual pursuit of self-interest leads to worse joint outcomes.

A tit-for-tat strategy was submitted by Anatol Rapaport. It is disarmingly simple: It cooperates on the first round with every opponent. Then it reciprocates whatever the opponent did in the previous round. An opponent's cooperation is rewarded with immediate cooperation. The tit-for-tat was not blindly cooperative, however: it met an opponent's competition with competition. Defection was punished with immediate defection.

Axelrod held a second tournament that attracted the eager submission of sixty-two strategies. All of the entrants knew the results of the first round—namely, that tit-for-tat had won. All had the opportunity to return to their blackboard, to adjust their mathematical algorithms and carry out further computer simulations, and to devise a strategy that could unseat the tit-for-tat. In this second tournament, once again the tit-for-tat prevailed. The tit-for-tat did not prevail, it is important to note, against all strategies. For example, your more sinister mind might have anticipated that a strategy that starts out competitively and always competes will have the upper hand against the tit-for-tat, because it establishes an advantage in the first round (of course, this strategy scores few points, and suffers profoundly, against other purely competitive strategies). Overall, however, tit-for-tat, so simple and cooperative in its *jen*-like design, achieved the highest outcomes against the society of different strategies in the tournament.

Why tit-for-tat? Three principles underlie the tit-for-tat and also underlie emotions like compassion, embarrassment, love, and awe, which promote the meaningful life. A first is what might be called cost-benefit reversal. Giving to others is costly. Devoting resources to others—food, affection, mating opportunities, protection— entails costs to the self. In the long run, generosity risks dangerous exploitation if it is directed at others who do not reciprocate in kind. The costs of giving constrain the tendency toward cooperation.

Built into the human organism, therefore, must be a set of mechanisms that reverse the cost-benefit analysis of giving. These mechanisms might prioritize the gains of others over those of the self, and transform others' gains into one's own. The tit-for-tat instantiates this principle of cost-benefit reversal. Its default setting is to cooperate, to benefit the other as well as the self. It is not envious; the tit-for-tat does not shift strategy as its partner's gains mount. And it forgives; it is willing to cooperate at the first cooperative action of its partner, even after long runs of mean-spirited defection.

The emotions that promote the meaningful life are organized according to an interest in the welfare of others. Compassion shifts the mind in ways that increase the likelihood of taking pleasure in the improved welfare of others. Awe shifts the very contents of our self-definition, away from the emphasis on personal desires and preferences and toward that which connects us to others. Neuro-chemicals (oxytocin) and regions of the nervous system related to these emotions promote trust and long-term devotion. We have been designed to care about things other than the gratification of desire and the maximizing of self-interest.

A second principle is what we might call the principle of reliable identification. This is clearly evident in the tit-for-tat—it is easy to read. There is no trickery to it, no Machivellian dissembling, no strategic misinformation. It would likely take only five to ten rounds against the tit-for-tat to make confident predictions about its future moves. Contrary to what you see on cable poker tournaments (where stone faces and inscrutability are the demeanor of the day), in the emergence of cooperative bonds transparency of benevolent intent is the wiser course. Cooperation is more likely to emerge and prosper when cooperative individuals can selectively interact with other good-natured individuals.

The implication is clear: Cooperation, kindness, and virtue are embodied in observable acts—facial muscle movements, brief vocalizations, ways of moving the hands or positioning the body, patterns of gaze activity—that are signals detectable to the ordinary eye. These outward signals of virtue, it further stands to reason, have involuntary elements that are not likely to be faked, and are likely to be put to use as people form intuitions about whom to trust and love and sacrifice for. This central premise—that for cooperation and goodness to emerge there must be outward signs of trustworthiness and cooperation—shapes the very design of the nonverbal signs of compassion, gratitude, and love. As science has begun to map the pro-social emotions in the body, new facial displays of embarrassment, shame, compassion, awe, love, and desire have been discovered. Studies of new modalities of communication, such as touch, have revealed that we can communicate gratitude, compassion, and love with a brief touch to the forearm. We are wired to detect benevolent intent in others in the moment-to-moment flow of the microinteractions of our daily living.

Finally, the tit-for-tat evokes cooperation in others—the principle of contagious cooperation. The tendency to cooperate and give can be readily exploited by individuals who are competitive and self-serving; nice guys do finish last in certain contexts. Kind individuals fare better, however, if they are able to evoke pro-social tendencies in others, thus prompting cooperative exchange. To the extent that goodness evokes beneficent responses in others, it should flourish.

Compassion, embarrassment, and awe are contagious at many different levels. Perceiving a person's smile, even below subliminal awareness, prompts the perceiver to feel good and to show shifts away from fight-flight physiology. Perhaps more remarkable are the feelings evoked in hearing of others' kindness—the swelling in the chest, goosebumps, and occasional tearing. Jonathan Haidt has called this state elevation, and he argues that we're wired to be inspired by hearing the good acts of others. Through touch, cooperation and kindness can spread across people and physical space within seconds. The emotions that promote the meaningful life are powerfully contagious, which increases their chance for propagation, and their encoding into our nervous systems and their ritualization into cultural practice.

We have now set the stage for our examination of emotions that promote high *jen* ratios and the meaningful life. We have reviewed the intellectual backdrop in which this work has taken place, which has assumed that emotions are disruptive, base tendencies, part of a human nature largely oriented toward the gratification of desire. We have considered the specifics of emotions that have been discovered in the past thirty years. We have learned that emotions serve as commitment devices, are embodied in our bodies, and shape moral judgment in systematic fashion. And in this past chapter we have sketched what kind of evolutionary environment might have given rise to emotions like compassion or gratitude, and what general principles these emotions abide by. We will now turn to scientific studies that illuminate this new swath of human design, and that will lend credence to Darwin's insight about the origins of human goodness: that it is rooted in our emotion, and that these social instincts may be stronger than those "of any other instinct or motive."

5

Embarrassment

O N JULY 2, 1860, Eadweard Muybridge boarded a stage-coach in San Francisco bound for St. Louis, Missouri, where he was to catch a train and make his way to Europe. There he would search for rare books to fill the shelves in the bookstore that he ran with his brother. In northeastern Texas, things went horribly. The driver of the stagecoach lost control and the coach careened down a hillside. Muybridge was hurled out of the boot of the coach, smashing his face against a tree, damaging a part of his frontal lobes that enables people to draw upon their emotions in making difficult decisions.

After six vague years in England, Muybridge returned to San Francisco. In 1872, he married Flora Shallcross Stone, twenty-one years his junior. While Muybridge was away on assignment for weeks on end, taking photographs of Yosemite and the Indian wars, Flora frequented fashionable theaters and restaurants with the dashing Major Harry Larkyns. Flora soon bore a baby boy. The little boy was more the source of uneasy suspicion than joy for Muybridge. Muybridge's concerns were quickly confirmed: He found a photo of the baby with "Little Harry" inscribed on the back. When the baby's nurse confirmed Muybridge's suspicions—that Harry Larkyns was the father of the baby—Muybridge was overwhelmed.

Eadweard Muybridge

He took a train to Calistoga, where Larkyns was working at the Yellow Jacket Ranch. Once at the ranch, Muybridge strode up to the front door and summoned Larkyns. When Larkyns arrived, Muybridge stated in matter-of-fact fashion: "Good evening, major, my name is Muybridge," at which time he raised his Smith & Wesson No. 2 six-shooter and shot Larkyns one inch below his left nipple. Larkyns grabbed his wound, ran through the house to his friends outside, and collapsed and died. A witness to the scene disarmed Muybridge and took him to the parlor, where Muybridge apologized to the women present for the "interruption."

At Muybridge's highly publicized trial, which produced an acquittal, several witnesses spoke of the changes the stagecoach accident had produced in his character. After the accident, he seemed like a different man—eccentric, remote, aloof, and cold. His speech and manner of dress were odd. He did not clean himself regularly. He cared little for social outings. He had difficulties keeping track of the contracts that financed his photography. And he demonstrated little or no modesty, no embarrassment at his eccentricities.

What does embarrassment have to do with incivility, remoteness, and murder? To find answers to these questions, I trained my eye in the frame-by-frame view of human social life inspired by Darwin and pioneered by Muybridge himself in his still photography. I slowed down the blur of two-second snippets of embarrassment and studied its fleeting elements—gaze shifts, head movements down, coy, compressed smiles, neck exposures, and glancing touches of the face. At the time I began my research, the display of embarrassment was thought to be a sign of confusion and thwarted intention. My research told a different story, about how these elements of embarrassment are the visible signals of an evolved force that brings people together during conflict and after breeches of the social contract, when relations are adrift, and aggressive inclinations perilously on the rise. This subtle display is a sign of our respect for others, our appreciation of their view of things, and our commitment to the moral and social order. I found that facial displays of embarrassment are evolved signals whose rudiments are observed in other species, and that the study of this seemingly inconsequential emotion offers a porthole onto the ethical brain, which in Muybridge's case had been destroyed in northeastern Texas over 140 years ago.

SLOW WORLD, FRAME BY FRAME

When Muybridge returned to California in 1866, brain damaged and a different man, he was swept up in a period of radical change. Space and time and the ordinary rhythms of human exchange were being annihilated by the new technologies, the steam engine, the railroad, the factory, and photography. Muybridge became a photographer of this modernization, this deconstruction of human social life.

Muybridge is best-known for his studies of animals in motion, an obsession that began with his photos of Leland Stanford's horses on his farm in Palo Alto. In a frenzied eighteen months at the University of Pennsylvania, Muybridge shot over 100,000 photos, capturing the frame-by-frame elements of people, often nude, walking,

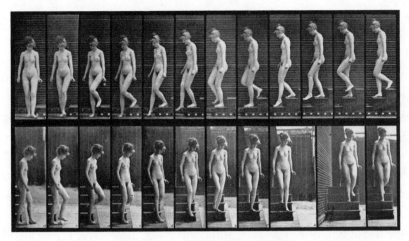

A body in motion by Muybridge

running, doing flips, jumping, throwing discs, descending stairs, and pouring water. He shot nude women throwing balls and feeding dogs, a legless boy getting into and out of a chair, cripples walking, and near-nude men doing rifle drills, laying bricks, and throwing seventy-five-pound rocks. The subjects' faces are typically turned away. They are lonely forms removed from the warm surround of other people.

In this frame-by-frame world, Muybridge revealed truths previously inaccessible to the human eye, truths about whether horses' hooves are all aloft when galloping, about the coordination of arms and legs during a simple stroll, about how the arms thrust backward after throwing a heavy object. Slow-motion scenes in film are similarly revelatory. In one scene in Martin Scorsese's *Raging Bull*, Robert De Niro, as the 1940s middleweight Jake LaMotta, first realizes his desire for a young teen while staring at her feet, splashing in a pool, foot by immersing foot, in hypnotic slow motion. On the ropes in a bloody championship fight, LaMotta and Sugar Ray Robinson dwell in intimate eye contact in a few slow-moving frames, realizing respect amid violence, through their swollen eyes, misshapen heads, and the stroboscopic blur of their punches.

For Darwin, the frame-by-frame world revealed how human facial expression traces back to the expressions of our primate relatives, and the selection pressures that have produced the human emotional repertoire. It is this frame-by-frame world to which Paul Ekman and Wallace Friesen devoted seven years, developing the Facial Action Coding System.

It was this frame-by-frame world that I entered into in 1990 as a post-doc in Paul Ekman's Human Interaction Laboratory. The lab is tucked away in a fog-bound, beige, two-story Victorian amid the Mies van der Rohe-style steel-and-glass UC San Francisco campus. My first task, of Talmudic proportions, was to master the Facial Action Coding System, which, as mentioned earlier, takes 100 hours of monastic, vision-blurring study. There is the manual itself, seventy dense pages, in which all visible facial actions are translated to specific action units, or AUs, and combinations of AUs. There is the instructional videotape, in which you watch Ekman move each individual facial muscle and then see important combinations, demonstrating the grammar of facial expression, the periodic chart—the skeptical, bemused, outer eyebrow raise (AU2), the delicate drawing together of the lips (AU8), the tragic raising of the inner eyebrow (AU1), the sulky lip curl (AU22), the hackle-raising tightening of the lower eyelid (AU7).

As I walked the streets from my apartment in San Francisco to the Human Interaction Laboratory, a frame-by-frame world exploded with AU12s (smiles), AU4s (frowns), AU5s (glares), AU9s (nose wrinkles), and AU29s (tongue protrusions). I began to see still, frozen vestiges of our evolutionary past in the crowded, unfolding world: flirtations between two teens waiting for a tram; festering anger between husband and wife stewing at a table in a café; radiant warmth in the shared gaze of a nine-month-old and her mother, lying on a picnic blanket. In these instances I began to see the products of millions of years of evolution, the traces of positive emotions that bind humans to one another.

Once, I was lacing up my hightops for a game of pickup basketball near some creaky, rusty swings. A tense mother was pushing her eight-year-old daughter on a swing. As the young girl swung by, her face was frozen in the tightened, lifted eyebrows, the remote eyes, and taut, elongated mouth that telegraphed chronic anxiety. As she returned to my field of vision with each backward swing, her face remained frozen in this expression, one faintly mirrored on her mother's face. In this thin slice of time, the lifetime of anxiety she faced was evident. Inspired by my in-lab and out-of-lab observations, I began to see the origins of the evolution of our ethical sense in brief displays of embarrassment.

THIN SLICES OF MORTIFICATION

My first project with Ekman, a rite of passage, really, was to code the facial actions of people being startled. The startle is a lightning-fast response that short-circuits whatever the individual is immersed in—reading a newspaper, snacking on a bagel, daydreaming of warm sand and a novel. That is the orienting function of the startle: it resets the individual's mind and physiology to attend to the source of the loud noise that has suddenly entered the individual's phenomenal field.

The startle response involves seven actions: a blink, cheek tighten, furrowed brow, lip stretch, neck tighten, and shoulder and head flinch, which blaze by in a 250-millisecond blur. Coding them is a form of torture, like watching a sky for shooting stars, knowing they're going to appear, and being asked to pinpoint the exact instant and place where they appeared and when and where they gracefully dissipated. Why was I devoting precious publish-or-perish time to coding the startle response? Wasn't there bigger game for me to set my sights on?

As it turns out, the magnitude of the 250-millisecond startle response is a telling indicator of a person's temperament, and in particular of the extent to which the person is anxious, reactive, and vigilant to threat and danger. People with intense startle responses, typically measured in terms of the intensity of their eye blink, experience more anxiety and dread. They are more tense and neurotic. They are more pessimistic about their prospects. The startle response is a good bet to capture a veteran's degree of post-traumatic stress disorder. If you're worried about moving in with someone who might be a bit too neurotic for your tastes (and this could be justified; neurotic individuals make for more difficult marriages), consider startling him and gathering a bit of data.

Sneak up on your beloved as he is settling into a glass of wine, and drop a heavy book on the counter next to him. If he shrieks, with arms flailing and wine glass flying, you have just witnessed a few telling seconds of his behavior that speak volumes to how he will handle the daily stresses and tribulations of life.

The participants I was coding—UC Berkeley undergraduates—sat alone in a room staring at a monitor. The experimenter asked the participant to relax and wait for the next task. The participant appeared to lapse into a dissociative state, drifting into thoughts about what the Oedipal complex really means, what the schizo-phrenic poet on the street corner was yelling about, or whether the day's hot temperature is another sign of global warming. And then—BAM!—an unexpected 120-decibel blast of white noise, as piercing as a pistol shot. There indeed was the startle response in all of its technicolor glory. People's faces clenched as they flinched uncontrollably, some almost lurching out of their chairs.

And then I noticed something unexpected. In the first frame after the startle response, people look purified, cleansed, as if their body and mind had been shut down for a second and then turned on—the orienting function of the startle. And then in the next frame their gaze shifted to the side. A knowing, abashed look washed over their faces. People looked as if they had been goosed, or whispered to of something lewd. And then a flicker of a nonver-bal display that Darwin had actually missed. Participants averted their gaze downwards, they turned their head and body away, they showed an awkward, self-conscious smile. Some blushed. Some touched their cheeks or noses with a finger or two.

Hastily I took videoclips of six of these participants to Ekman in his office downstairs. As we reviewed these two- or 3-second snip-pets, Ekman shook his head, first side to side, and then up and down in brief staccato bursts, smiling. He had seen these expres-sions in the Foré of New Guinea. He knew the contours of an emo-tional signal whose evolutionary story one could tell. He turned to me with a gleam in his eye. There was a signal of emotion there, one that the field had ignored.

CHARTING CRIMSON FACES

My first step was to embarrass people, a task that has given license to a more mischievous side of researchers' imaginations. To produce embarrassment in a laboratory, researchers have had college students suck on pacifiers in front of friends. Students have modeled bathing suits for an experimenter, taking notes, clipboard in hand. Young children have been overpraised by adults avidly snapping photographs of the child (prior to eighteen months, children absorb this flood of attention with the aplomb that the removal of a bib might prompt; after eighteen months of age, they show embarrassment). In perhaps the most mortifying experiment, participants had to sing Morris Albert's song "Feelings" using dramatic hand gestures. At a later date, they returned to the lab to watch a film clip with other students, which turned out to be of their performance of that cloying song.

Before I began my study of embarrassment, Paul Rozin of the University of Pennsylvania recommended two new paradigms, which I never had the temerity to employ. In a first, participants would ride alone in an elevator. Just prior to a group of people entering into the elevator at the next stop, I would surreptitiously plant the smell of a fart, and watch the participant squirm with embarrassment as new individuals entered with eyebrows raised in disbelief. In a second, I would give the participant a handkerchief filled with gobs of mucous. Then another individual, planted by me in the experiment, would ask the participant for the use of the handkerchief.

As appealing as these methods were to my more theatrical inclinations, my production of embarrassment needed to be constrained in certain ways. I had to choose a task in which participants' heads were relatively stationary so that I could code their facial actions in frame-by-frame analysis (head and body movements can reduce the visible traces of facial muscle actions to impressionistic blurs). I had to ensure that participants did not move their facial muscles after the embarrassing episode, so I could isolate the actions only accompanying embarrassment. In light of these constraints, I had individuals follow muscle-by-muscle instructions to achieve a difficult facial expression, guided by a martinet of an experimenter, all the while being videotaped. The instructions were as follows (try it if no one is looking):

1. Raise your eyebrows
2. Close one eye
3. Pucker your lips
4. Puff out your cheeks

The experimenter quickly noted, with drill sergeant precision, participants' deviations from the instructions ("keep your eyebrows raised"; "your eyes are fluttering, please just keep one eye closed"; "now close your mouth, and don't press your lips together, pucker them"; "remember to puff out your cheeks, and don't stick out your tongue"). Typically, after a valiant, thirty-second struggle, participants achieved the expression and then were asked to hold it for ten seconds. As participants' facial muscles quivered and they tried to hold their smiles at bay, they showed visible signs, a furtive glance askance, of imagining what their appearance was, permanently recorded on videotape: They looked like Popeye, drunk on ale, puckering up for a smooch from Olive Oyl, sure to be rebuffed. They were part of some weird joke or an act of absurdist theater. After ten seconds of this pose, participants were asked to rest. It was in the milliseconds after resting that I saw my quarry: the embarrassment display.

With these videos in hand, I spent much of a summer in the coding room of Ekman's lab, with its cream-colored walls and drawers bursting with electrical plugs, wires, and videotapes. Each fifteen-second snippet of behavior required about half an hour to code, as I charted each twenty-millisecond shift in gaze and discerned the specific muscle actions that define the awkward, embarrassed smile. At the time, most scientists assumed that the display of embarrassment was a jumble of confused actions. In real time my participants did appear rather shaken, uncertain, and disorganized.

Yet with careful frame-by-frame analysis a different picture emerged, and one in line with Darwin-inspired analyses of emotional displays as involuntary, truthful signs of our commitments to partic-

A display of embarrassment

ular courses of actions. Our facial expression of anger, for example, signals to others likely aggressive actions, and prompts actions in others that prevent costly aggressive encounters. Within this school of thought, emotional displays are highly coordinated, stereotyped patterns of behavior, honed by thousands of generations of evolution and the beneficial effects displays have on social interactions. Evolved displays unfold briefly, typically between two and three seconds. The brevity of emotional displays is, in part, due to limits on the time that certain facial muscles can fire. Emotional displays are brief, as well, because of the pressing needs facial expressions are attuned to—the approaching predator, the child catapulting toward danger, the flickering signs of interest shown by a potential mate amid many suitors. Involuntary displays of emotion have different temporal dynamics than nonemotional displays: They are gradual in their onset and offset. More voluntary displays, in contrast, like polite smiles, pouts, dramatic glares, or provocative puckers, can come on the face in milliseconds, and remain on the face for minutes, hours, days, or, for some regrettable souls, a lifetime.

What I charted in the elements of the embarrassment display was a fleeting but highly coordinated two- to three-second signal. First the participant's eyes shot down within .75 seconds after finishing the pose of the awkward face. Then the individual turned his head to the side, typically leftward, and down within the next .5 seconds, exposing the neck. Contained within this head motion down and to the left was a smile, which typically lasted about two seconds. At the onset and offset of this smile, like bookends, were other facial actions in the mouth, smile controls: lip sucks, lip presses, lip puckers. And while the person's head was down and to the left a few curious actions: the person looked up two to three times with furtive glances, and the person often touched his or her face. This three-second snippet of behavior was not some bedlam of confused actions; it had the timing, patterning, and contour of an evolved signal, coordinated, brief, and smooth in its onset and offset.

BARED-TEETH GRINS AND NODDING GULLS

To understand the deeper meaning of facial displays, like smiles, sneers, tongue flicks, or eyebrow flashes, researchers can do what Darwin had pioneered: turn to the displays of other species. By looking to other animals we discern the deeper forces that have produced many of the displays that we observe today. We learn about the contexts in which displays emerged—for example, when sharing food, fighting a rival, engaging in rough-and-tumble play, or protecting vulnerable offspring. We learn how displays are really the tip of the iceberg of more complex behavioral systems, such as eating, breastfeeding, attack, or defense.

Consider the kiss or, in Facial Action Coding System terms, the simple lip pucker and lip funnel (AUs 18 and 22), and, in more lascivious moments, the tongue protrusion (AU29). It is well known that people kiss differently in different cultures. In some cultures, kissing in public is rare or nonexistent, as with certain Amazonian tribes or the people of Somalia. There are different kisses for friends, political officials, children, and romantic partners. A visit to kissingsite.com will tell you there are thirteen kinds of romantic kisses, from the suck on the chin to quietly sharing breath. And there are, of course, individual extremes: One Italian couple kissed continuously for 31 hours, 18 minutes, and 33 seconds. In 1991 Alfred Wolfram kissed 8,001 people in eight hours at a Renaissance fair in Minnesota.

This sublime variety might seduce you into thinking that the kiss is a cultural artifact, like the peace sign, BlackBerry, fork, or necktie: Some cultures have it and others do not, and members of different cultures vary widely in their uses of the artifact. In fact, certain anthropologists have made such an argument about the kiss. Based on the absence of portrayals of kisses in cave paintings, they have argued that humans invented the kiss around 1500 BC, and that it spread from India westward. It was widely popularized by the Romans, who integrating kissing into numerous public rituals, such as kissing the ring of the emperor or other sacred objects.

This argument ignores what we learn by cross-species comparisons of the kiss. Our primate predecessors premasticate food to make it more digestible for the young, and deliver this softened caloric mass to the young with a kiss. The same has been documented by Irenäus Eibl-Eibesfeldt of preindustrial human cultures. Parents will chew up food and pass it on to their young offspring, mouth to mouth. Food sharing, then, is the original evolutionary context of the kiss. Primates, in their cooperative ways, have extended this rewarding display to acts of affiliation: They rely on lip smacks and pouts as signals to prompt others to come closer (see photo below). The human kiss has its roots in the food sharing of our close primate relatives.

What evolutionary forces gave rise to gaze aversion, head turns and face touches, and that coy smile of embarrassment? I found answers from studies of appeasement and reconciliation processes in nonhuman primates. Frans de Waal has devoted thousands of hours to the study of what different primates—macaques, chim-

A primate pouting display

panzees, and bonobos—do following aggressive encounters. Prior to this work, the unquestioned assumption was captured in the dispersal hypothesis: Following an aggressive encounter, combatants would move away from each other as far as possible, a safe, self-preserving, and adaptive thing to do.

Yet de Waal observed the opposite pattern of behavior. Instead of moving away from one another after conflict, the primates he was observing were more likely to spend time in the presence of one another. This would make sense for species that are so dependent upon one another to accomplish the basic tasks of survival and reproduction. With more careful observations, de Waal discovered how primates reconcile during conflict, and reestablish cooperative relations. In the midst of conflict or aggression, the subordinate or defeated animal first approaches and engages in submissive behaviors, such as bared-teeth displays, head bowing and bobbing, and grunts. These actions quickly prompt affiliative grooming, physical contact, and mutual embraces, reconciling the warring parties. In nonhuman primates, these reconciliation processes transform life-threatening conflicts into affectionate, backslapping embraces within seconds.

When I reviewed forty studies of appeasement and reconciliation processes across species, from blue-footed boobies to 4,500-pound elephant seals, the evolutionary origins of embarrassment became apparent: It is a display that reconciles, that brings people together in contexts of distance and likely aggression.

Let's take it behavior by behavior, in the Darwinian fashion. Gaze aversion is a cut-off behavior. Extended eye contact signals continue what you're doing; gaze aversion acts like a red light, terminating what has been happening. Our embarrassed participants, by quickly averting their gaze, were exiting the previous situation. They were signaling an end to the situation for obvious reasons: embarrassment follows actions—including social gaffes, identity confusion (forgetting someone's name), privacy violations (walking in on someone in a bathroom stall), and the loss of body control (the prosaic fart or stumble)—that sully our reputations and jeopardize our social standing.

What about those head turns and head movements down? Various species, including pigs, rabbits, pigeons, doves, Japanese quail, loons, and salamanders, resort to head movements down, head turns, head bobs, and constricted posture to appease. These actions shrink the size of the organism, and expose areas of vulnerability (the neck and jugular vein, in the case of human embarrassment). These actions signal weakness. Darwin himself arrived at a similar analysis of the shoulder shrug, which typically accompanies the recognition of ignorance (or intellectual weakness) and appears as the opposite of the postural expansion of dominance. At the heart of the embarrassment display, as in other species' appeasement behaviors, is weakness, humility, and modesty.

The embarrassed smile has a simple story with a subtle twist. The smile originates in the fear grimace or bared-teeth grin of nonhuman primates. Go to a zoo and watch the chimps or macaques, and you'll see subordinate individuals grin like fools as they approach dominant peers. Yet the embarrassed smile is more than just a smile; it has accompanying muscle actions in the mouth that alter the appearance of the smile. The most frequent one is the lip press, a sign of inhibition. When people encounter strangers in the street they often greet each other with this modest smile. Just as common are lip puckers, a faint kiss gracing the embarrassed smile as it unfolds during its two- to three-second attempt to make peace. Within reconciliation, many primates turn to sexual displays—rump presentations, genital touch and contact, and sexual mounting. While humans are not so bawdy in how they short-circuit aggression, we do show signs of affection—subtle lip puckers—in our embarrassment, to warm hearts and bring others closer. This explains why embarrassment displays and the coy smile are put to good use during flirtation and courtship.

The face touch may be the most mysterious element of embarrassment. Several primates cover their faces when appeasing. Even the rabbit rubs its nose with its paws when appeasing. Face touching in humans has many functions. Some acts of face touching act as self-soothing (the repetitive stroking of hair in the back of the head). Other face touches are iconic (the tragic rub of the inner eye; the flirtatious hair flick, which expands the coif to peacock-tail proportions). Certain face touches seem to act like the curtains on a stage, closing up one act of the social drama and ushering in the next. A psychoanalyst has even argued that we face-touch to remind ourselves that we exist, in the midst of social exchanges where our sense of self feels to be drifting away.

A clue to the origins of face touching in human embarrassment came from one participant from the original startle study. After she had been startled, she pulled her head into a shoulder shrug, and up went the hand, as if it was timed to deflect an aggressive blow. Some face touches (for example, covering the eyes) signal the exiting of the situation; others seem to be the residual actions of defensive postures. An element of embarrassment is self-defense.

In turning to other species' appeasement displays, the social forces that have shaped this display during the tens of millions of years of primate evolution were there to see. This simple display brought together signals of inhibition, weakness, modesty, sexual allure, and defense all woven together in a two- or three-second display. The mission of the display is to make peace, to prevent conflict and costly aggression, and to bring people closer together, to reestablish cooperative bonds. We may feel alienated, flawed, alone, and exposed when embarrassed, but our experience and display of this complex emotion is a wellspring of forgiveness and reconciliation. The complement would also prove to be true: The absence of embarrassment is a sign of abandoning the social contract.

EVANESCENT SIGNS OF MORAL COMMITMENT

Imagine that our most intimate relationships were arranged like speed dating. You are allowed one question to ask of others to figure out who will become lifelong friends, spouses, and work colleagues. What question would you ask? Do you call your mother regularly? How do you treat your cat? Have you ever thrown your back out trying to avoid stepping on an ant?

This thought experiment may sound absurd but in point of fact has clear parallels in analyses of the evolutionary origins of cooperation. Being good to others has many costs, and exposes the individual to exploitation by those who are less generous. Given the costs and risks of cooperation, we are on the hunt for subtle, unspoken signs of integrity, honesty, kindness, and trustworthiness.

In this strange speed dating moral universe, I would ask people to tell me of their last embarrassing experience. I would then focus my eyes and carefully watch embarrassment ripple across their faces. Why put stock in an emotion so closely associated with the seemingly superficial aspects of social life—politeness, manners, and social conventions regarding the exchanges between strangers? Because the elements of the embarrassment are fleeting statements the individual makes about his or her respect for the judgment of others. Embarrassment reveals how much the individual cares about the rules that bind us to one another. Gaze aversion, head turns to the side and down, the coy smile, and the occasional face touch are perhaps the most potent nonverbal clues we have to an individual's commitment to the moral order. These nonverbal cues, in the words of sociologist Erving Goffman, are "acts of devotion . . . in which an actor celebrates and confirms his relation to a recipient."

One way to test this hypothesis—that embarrassment displays are evanescent signs of moral commitment—would be to study moral heroes, and look to see whether they show extraordinary embarrassment and modesty. That is, is the modesty, deference, and respect they have cultivated seen in their everyday visage? One cannot help but be struck by the deep modesty evident in the smiles of people such as Gandhi or the Dalai Lama, which show elements of embarrassment that I documented—gaze aversion, lip presses, and smile controls.

I chose to study the other end of the continuum—people prone to violence. My thesis was simple: To the extent that embarrassment displays reflect respect for others and a commitment to the moral order, the relative absence of embarrassment should be accompanied by the tendency to act in antisocial ways, the most extreme being violence. In a first study to test this hypothesis, I concentrated on young boys prone to violence, known in clinical science as externalizers (they externalize their inner turmoil by acting out aggressively). These are boys who fight, bully, steal, burn things, and

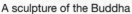

A sculpture of the Buddha   Gandhi   His Holiness
the Dalai Lama

vandalize on a routine basis. I observed ten-year-olds while they were taking a two-minute interactive IQ test, attempting to answer questions whose answers you'd find in an encyclopedia ("What is a barometer?" "Who was Charles Darwin?"). The test is designed to produce some failure in all children. All the boys in this study responded emotionally to these academic misfires, glaring in anger, showing the tightened brow of anxiety, or, most typically, showing the now familiar signs of embarrassment. Consistent with my moral commitment hypothesis, the well-adjusted boys showed the most embarrassment, and in fact this was their dominant response to the test. They in effect were displaying concern over their performance, and perhaps a deeper respect for the institution of education. The externalizing boys, in contrast, showed little or no embarrassment. Instead, these boys erupted with occasional facial displays of anger (one boy gave the finger to the camera when the experimenter momentarily had to leave the testing room). The fleeting, subtle embarrassment display is a strong index of our commitment to the social-moral order and the greater good.

Neuroscientist James Blair has followed up on this work on embarrassment and violence by studying "acquired sociopathy," that is, antisocial tendencies brought on by brain trauma. One such patient, J. S., was an electrical engineer. One day in his mid-fifties he collapsed and lost consciousness. During his recovery in a hospital, he was known for his outrageous outbursts. He threw furniture at other patients. He pushed a wheelchair-bound patient around at roller-coaster speed and with hairpin shifts in direction despite her shrieks of terror. He groped female nurses on a routine basis and, on another occasion, bodysurfed on a gurney through the hallways of the hospital.

In Blair's research, J. S. demonstrated a normal ability to learn, to recognize faces, and to identify whether faces were male or female. He showed normal physiological reactions to a clap and the utterance of his name. He could provide normal explanations of protagonists' behavior briefly described in vignettes, suggesting that he did not suffer from some general deficit in understanding others' mental states.

What J. S. proved to be deficient in was embarrassment. In one task, he was asked to attribute emotions to hypothetical characters in various situations. Some were ones of happiness—a person wins an award. Others were of sadness—the protagonist loses a job. Still others were about embarrassment—a protagonist slips in a café and falls prostrate in the middle of some tables. J. S. was near-perfect in his ability to attribute feelings of happiness and sadness to the fictional characters; he could reason about the emotions of gains and losses. He was utterly incompetent in his attributions of embarrassment.

Blair also explored J. S.'s reactions to slides portraying anger and disgust expressions—the kinds of expressions that often signal disapproval and trigger our embarrassment. How did J. S. respond to these signs of moral disapproval? He had great difficulty identifying the emotions conveyed. Unlike comparison participants, he also failed to show a skin-conductance response—the release of sweat from tiny glands underneath the skin of the fingers. His body was not wired to respond to the judgments of others.

J. S. had damaged the orbitofrontal cortex, a region of the frontal lobes that is adjacent to the jagged, bony ridges of the skull's openings for the eyes. This region of the brain is often damaged in falls and bike and motorcycle accidents, as the brain jostles during the fall and is carved up by the bony backside of the eye sockets. This damage had left J. S.'s reasoning processes intact, but it had short-circuited his capacity for embarrassment. In actuality, he had lost something much larger: his ability to appease, reconcile, forgive, and participate in the social-moral order. More in-depth studies of this region of the brain would tell us what might have changed in Eadweard Muybridge the fateful day he was thrown headfirst into a tree.

MUYBRIDGE'S IMMODEST BRAIN

When Eadweard Muybridge regained consciousness after his injury, he felt strange. He had no sense of smell or taste. He had double vision. In his own muted words, he had "confused ideas." Most likely those confused ideas centered upon a new disconnect to others, a sudden blindness to the rich web of conventions and subtle acts of cooperation that bind people to one another.

Like J. S., Muybridge had damaged his oribitofrontal cortex, which might be thought of as a command center for the moral sentiments. Anatomically, the orbitofrontal cortex receives information from the amygdala, a small, almond-shaped part of the midbrain, which provides a millisecond, unconscious assessment of whether objects are good or bad. It receives information from the cingulate cortex, which is involved in assessments of pain and harm. Soft, velvety touch to the arm activates the orbitofrontal cortex, suggesting that this portion of the brain tracks physical contact between people so central to the currency of gratitude and compassion and the formation of intimate and egalitarian bonds. It receives information from the vagus nerve, which is activated during our experience of compassion.

Remarkably, damage to the orbitofrontal cortex does not impair language, memory, or sensory processing, as Blair's study of J. S. revealed. Patients who damage these regions speak with the fluency that would please any grammarian, and the cogence that would satisfy the most persnickety of logicians. Cold reason remains intact. But damage to the orbitofrontal cortex does tend to turn individuals into impulsive, everyday psychopaths.

We know this from case studies of people who have damaged that region of the brain. The most famous is Phineas Gage, who accidentally blew a thirteen-pound tamping rod through his skull while

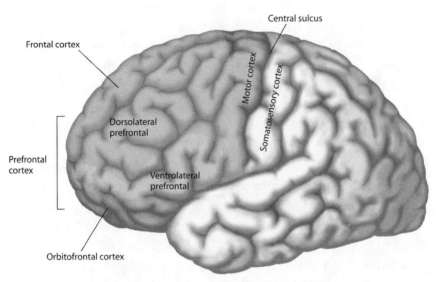

Central sulcus

Frontal cortex

Motor cortex

Somatosensory cortex

Dorsolateral
prefrontal

Prefrontal
cortex

Ventrolateral
prefrontal

Orbitofrontal cortex

A side view of the brain showing the orbitofrontal cortex.

working on the Rutland and Burlington Rail Road in Vermont. The doctor who cared for Gage, John Harlow, offered one of the few recorded observations about Gage, who, prior to the accident, was uniformly considered a considerate, reliable, upstanding man: "He is fitful, irreverent, indulging at times in the grossest profanity (which was not previously his custom), manifesting but little deference for his fellows, impatient of restraint or advice when it conflicts with his desires."

In research with Jennifer Beer and Robert Knight on orbitofrontal patients, we sought to document that these patients, so skilled in the tasks of cold reasoning, have lost the art of embarrassment. They have lost the ability to appease, to reconcile, and signal their concern for others. In the study, our participants navigated a veritable obstacle course of embarrassing traps and hurdles. First, they disclosed personal experiences to a relative stranger—an exercise fraught with the possibility of being inappropriately intimate. Participants then teased an attractive female experimenter whom they had just met. They did so by making up a nickname and a provocative story for that person. Finally, patients were presented with slides of different facial expressions of emotions, including one of embarrassment—a trigger of recon-ciliation and forgiveness.

Our patients barreled through these tasks with the immodest impulse of the wild-eyed, street-corner psychopath. In the emo-tional disclosure task, comparison participants talked about being embarrassed at forgetting someone's name or not understanding the punchline of a joke. Orbitofrontal patients, in contrast, recounted experiences that were often sexual and more suitable to a therapy session than an interaction with a stranger. They were unconstrained by the anticipation of embarrassment at having crossed the bound-aries of intimacy. One patient's account of embarrassment to his new acquaintance, the experimenter: "I was embarrassed when I was dis-covered in a store's dressing room with my girlfriend."

When teasing the stranger, the orbitofrontal patients did so in inappropriate and often lewd fashion. The nicknames they devised always contained sexual innuendo directed at the experi-menter. One joked about what he and the experimenter might really get down to if given the chance. Unlike the comparison par-ticipants, the orbitofrontal patients showed no signs of embarrass-ment when teasing, even though their provocative efforts were often quite outlandish.

Finally, in judging the emotions of others, our orbitofrontal patients were inept at identifying embarrassment from photos, although they were quite skilled at judging other facial expressions, for example those of happiness, amusement, or surprise. They resembled psychopaths, who prove to be unresponsive to the signs of suffering in others.

Embarrassment warns us of immoral acts and prevents us from mistakes that unsettle social harmony. It signals our sense of wrong-doing and our respect for the judgments of others. It provokes ordi-nary acts of forgiveness and reconciliation, without which it would be a dog-eat-dog world. Orbitofrontal patients, fully capable in the realm of reason, have lost this art of embarrassment. They have lost the subtle ethic of modesty.

AN ETHIC OF MODESTY

Philosophers turn to metaphors to describe the moral sentiments, and those metaphors often center upon animating natural forces that unite humans in common cause. For the British Enlightenment philosophers, moral sentiments like sympathy made up an invisible force field, binding individuals to one another. For the Chinese philosopher Lao Tzu, the Tao, or way of virtue, is like water, noncompetitive but touching all. Embarrassment is like an ocean wave: It throws you and those near you into the earth, but you come up embracing and laughing.

The simple elements of the embarrassment display I had documented and traced back to other species' appeasement and reconciliation processes—the gaze aversion, head movements down, awkward smiles, and face touches—are a language of cooperation; they are the unspoken ethic of modesty. With these fleeting displays of deference, we preempt conflicts. We navigate conflict-laden situations (watch how regularly people display embarrassment when in close physical spaces, when negotiating the turn-taking of everyday conversations, or when sharing food). We express gratitude and appreciation. We quickly extricate embarrassed souls from their momentary predicaments with deflections of attention or face-saving parodies of the mishap.

Embarrassment is the foundation of an ethic of modesty. The display of embarrassment converts events that go into the denominator of the *jen* ratio (social gaffes, offensive remarks, violations of privacy) and transforms them into opportunities for reconciliation and forgiveness (experiences in the numerator of the *jen* ratio). It is in these in-the-moment acts of deference that we honor others, and in so doing, become strong. It is often when tender and weak that we are alive, and full of *jen*. In the words of Lao Tzu:

When man is born, he is tender and weak
At death, he is stiff and hard
All things, the grass as well as trees, are tender and subtle while alive
When dead, they are withered and dried.

Therefore the stiff and the hard are companions of death
The tender and weak are the companions of life
If the tree is stiff, it will break
The strong and the great are inferior, while the tender and the weak are
superior.

6

Smile

THE SETI PROJECT is the largest in the world devoted to communicating with intelligent life forms outside of those on Earth. A branch of SETI has brought together anthropologists, mathematicians, physicists, and media and communication experts to solve an intriguing problem: Which symbols should we send out into the infinite expanse of the universe to communicate the altruistic capacities of the human species? Assuming other intelligent life-forms emerged in similar carbon-based chemical processes as we did, how might we, given one shot, communicate our capacity for good to other intelligent minds? The yin-yang symbol? An image of a baby, with round eyes, small mouth, and minuscule mound of a chin? Perhaps, instead, we should rely on sound, given our powerful capacity to communicate vocally. How about the perfect laugh, a soothing sigh, a meditative oohmmmm, or coos between infant and parent?

The question that SETI scholars are debating mirrors one at the heart of this chapter: As our hominid predecessors increasingly lived and worked in close proximity with one another, gathering and distributing plants, fruits, and seeds, sharing the meat of a kill, tending to the needs of vulnerable offspring, moving through gatherings of potential mates and vigilant rivals, what behaviors allowed them to navigate such conflict-rife contexts in cooperative fashion? The classical Greeks had their own answer, one that will anticipate the theme of this chapter—the smile.

As Angus Trumble details in *A Brief History of the Smile*, in the third to fifth centuries BC, Greek artisans began sculpting the Kouros, a life-sized sculpture that has been found throughout mainland Greece, Asia Minor, and islands in the Aegean Sea. It is a sublimely dynamic sculpture, with upright posture, left foot moving forward, and hands gently clenched in determination. The most captivating aspect of the Kouros, though, is its smile (see figure below). It is at once modest, poised, expectant, and brimming with contained delight.

In its heyday, the Kouros served as an all-purpose symbol of goodness. It was commonly placed in ceremonial settings as an offering to the gods, to communicate reverence toward the higher powers that controlled the quirks of fate on the earthly ground of Greek life. It was a common presence at funerals, no doubt for those well-off enough to afford such memorializing, serving as an image of the deceased and as a symbol of the gods who would protect the soul of the deceased. For the Greeks, the Kouros represented the soul embodied in the human form.

The Kouros

Evolutionary analysis will tell a similar story about the human smile. In evolution's toolbox of adaptations that promote cooperation, the smile is perhaps the most potent tool. The smile is visible from hundreds of feet. It triggers, science has discovered, activation in reward centers of the brain. It soothes the stress-related physiology of smiler and perceiver alike. The smile smoothes the rough edges of our social life, creating a medium of benevolent exchange. The right kind of smile brings the good in others to completion. It is one of the first acts of *jen* in primate evolution.

At stake in our evolutionary analysis of the smile are answers to two questions. The first is straightforward, but has proven to be a surprisingly prickly source of controversy: What does the smile mean? People smile in almost every imaginable context: seeing a loved one, being sentenced to prison, enjoying ice cream and the appalling cooking of a dear friend, hearing that one is pregnant and receiving dire medical news, winning lotteries and losing Olympic competitions. The English language possesses a few words for smiles—"smile," "grin," "smirk," "beam"—really a paucity of concepts that masks the rich complexity of the realm of smiles. A better understanding of what the smile means will be found by turning to facial anatomy and evolutionary analysis.

A deeper question, however, is at play in our search for the origins of the smile: What are the roots of human happiness? If the right kind of smile is synonymous with happiness, which intuition and dozens of scientific studies suggest is the case, then our search back in time for the social contexts in which the smile emerged is really a search back in time for the origins of human happiness. And this journey would begin with Charles Darwin's intuitions, and end in studies of smile-like behavior in our more egalitarian primate relatives.

MISLED BY THE LAUGHTER OF CHILDREN

Sometimes vivid images produced by careful observation lead us astray. Such was the case in Charles Darwin's analysis of the smile. Darwin kept detailed recordings of the development of the emotional lives of his children. In writing about the emergence of laughter, he discerned a reliable pattern. At around fifty days his children would begin to smile. Gradually, with age, in similar contexts such as tickling, which he resorted to as scientist and devoted father, he would see, about two months later, the rudimentary signs of laughter—"little bleating noises"—that systematically were released during exhalation.

From these transfixing observations, Darwin arrived at his thesis about the smile: that it is the first trace of the laugh. Given this assumption, he then answered—rightly, I believe—the question of the morphological origins of the smile. Why does the smile take its characteristic form of lips retracted upward and occasionally to the side? Why do we not signal a sense of amusement with an eyebrow flash, a cheek flicker or nostril flare, or any of the other thousands of possible configurations of facial muscles? Darwin's answer is found in two claims. First, a nod to the principle of antithesis: We smile as a public offering of high spirits because the shape of the smile, with its curved movements upward, is the antithesis of the tightened lips, the lip corners pulled down, the bared teeth, of anger. The smile signals the antithetical state of its opposite expression, that of anger. The second observation is in keeping with Darwin's analysis of the physical actions that facial expression are part of: The smile's retraction of the mouth corners up, and occasionally, sideways, enables the kinds of exhalation and vocalizations seen in laughter.

Darwin's thesis, then, is that the smile is the first stage of the laugh, the larva to the butterfly, the acorn to the oak tree. There is something deeply satisfying in this view. Perhaps the Greeks had it right, that there are indeed two swaths of human emotional life: the tragic realm, a serious, fate-altering spectrum of emotions like anger, fear, and sadness, associated with tragic losses, threats, and injustices; and a comedic realm, defined by playful, lighthearted emotion grounded in laughter. Perhaps all of our positive states—enthusiasm, hope, gratitude, love, awe—originate in our ability to take alternative perspectives upon our current state of affairs: a prerequisite of the laugh.

Parsimonious and pleasing as this may be, it's wrong. When primatologist Signe Preuschoft put Darwin's smile-as-laughter thesis to the test by examining when various nonhuman primates show smilelike and laughterlike displays, she found that these two displays occur in much different social contexts, and toward much different ends. The smile and the laugh originate in distinct slices of early primate life, and have subsequently followed separate evolutionary trajectories as they worked their way into the human emotional repertoire and our nervous systems.

SILENT BARED-TEETH AND RELAXED OPEN-MOUTH DISPLAYS

In her careful observations of primates, in particular several different macaque species, Preuschoft has catalogued numerous displays that convey affiliative, cooperative intent. These include pout faces and lip smacks (which Darwin wrote about—see chapter 2)—no doubt predecessors to the succor-seeking sulking we see in three-year-olds, and of course, the kiss. The most common affiliation-

The chimp on the left displays a classic silent bared-teeth expression, the predecessor to the human smile. The chimp on the right, collapsed in a bout of tickling, is showing the play face, the predecessor to human laughter.

seeking displays in primates, and most central to our understanding of human smiling and laughter, are the silent bared-teeth display and the relaxed open-mouth display (see previous pictures).

Across species, primates resort to the silent bared-teeth display to appease and to signal submissiveness, weakness, and social fear in contexts in which the likelihood of conflict and aggression is high, for example when nearing dominant primates. The silent bared-teeth display is most typically seen in submissive primates and is usually accompanied by inhibited posture, protective body movements such as shoulder and neck tightening, or hands hovering around the face for obvious defensive purposes. Thankfully, this display often short-circuits aggression, triggering reconciliation in the dominant monkey—affiliative grooming and embracing.

In humans, the silent bared-teeth display is evident in our deferential smile, which signals thoughtful, at times fearful, attention to the concerns of others (see pictures below). This smile involves the activity of two muscles: the zygomatic major, which pulls the lip corners upward, and the risorius, which pulls the lower lip sideways. I first encountered the deferential smile empirically in an early study of teasing, in which I had two high- and two low-status fraternity brothers tease one another. As they ripped into each other in the

The deferential smile looks different from other kinds of smiles due to the movement of the lower lip sideways (through the contraction of the risorius muscle). Why only British examples of the deferential smile? My own intuition, having lived in England twice, is that the British have a more sophisticated system revolving around politeness and deference than Americans do, and as a result have cultivated a more deferential smile. More rigorous science is needed to ascertain whether indeed the Brits are more likely to smile in deferential fashion than members of other cultures, or whether this is just a stereotype embodied only in Claymation figures and heirs to the throne.

profane gutter language of young men living together in tight quarters, the low-status guys were ten times more likely to show deferential smiles. A good time was had by all, but it was the low-status guys who signaled their subordinate positions with this smile.

The relaxed open-mouth display, in contrast, is observed, Preuschoft notes, in fewer primate species. It is accompanied by panting and staccato breathing, and on occasion bursts of grunt- or howl-like vocalizations and boisterous body movements. Quite clearly, the relaxed open-mouth display is the primate predecessor to the human laugh. Importantly, Preuschoft has found that the relaxed open-mouth display occurs in a radically different set of social contexts than those associated with the silent bared-teeth display: It precedes and accompanies the pyrotechnics of primate play—chasing, nuzzling, gnawing, rough-and-tumble somersaults and cavorting in the branches of trees.

Preuschoft's analysis of these two primate displays makes it difficult, even for the ardently faithful, to continue entertaining Darwin's hypothesis that the smile is the first stage of the laugh. No longer tenable as well is the pleasing inference that our capacity for play is the most rudimentary element of positive emotion. Instead, we must conclude that smiling and laughter have distinct evolutionary origins. The smile emerged to facilitate cooperative and affiliative proximity. The laugh emerged to promote play and levity. They are tokens of different swaths of positive emotion, and different facets of the meaningful life.

A VOCABULARY OF SMILES

During the summer following my freshman year in college, I decided to teach myself classical guitar while living at home in Penryn, California, a tiny rural backwater named after an island in Wales. Two weeks into thick-fingered attempts at "Classical Gas," my mother had had enough. A week later I found myself donning the brown polyester and golden arches insignia of the McDonald's uniform, serving burgers, fries, Chicken McNuggets, and gooey sundaes to sunburned revelers on their way to underaged drinking and debauchery at the rocky rivers in the foothills of the Sierras or the noisy waterskiing lakes. Each and every day at 11:10 AM a middle-aged man arrived, strode to the counter in shoes that made a strange clicking sound, and, with somber brown eyes and Lincolnesque sideburns, placed the same order: four plain hamburgers, with nothing on the gray patties and buns that dissolved upon touch, and a cup of black coffee, which I had to refill a dozen or so times in the span of the thirty-six minutes he reliably took to finish his lunch. He became a Sisyphus-like commentary on my fate: the minimum-wage undermining of my musical career and the missed opportunities for summertime revelry. My manager, a good-hearted, optimistic soul, recognized my deep despair and offered managerial guidance straight out of some McDonald's handbook: just smile. I felt deeply oppressed, filling the regular customer's Styrofoam cup with another round of coffee, smiling as I delivered his cup of joe.

I can assure you that I was not smiling the smile that evolution has produced, and which we will soon dissect, and which promotes goodwill between individuals. Much more likely, I was emitting the service industry smile, the one that signals that the customer is always right, that the sale should always come first. Sociologist Arlie Hochschild has argued that this smile is part of the emotional labor required of so many service-oriented jobs and the tip of the iceberg of alienation from the fruits of human labor. Research shows that when workers smile in the service industry, for example when greeting customers at a 7-11 counter, customers are more satisfied and actually more likely to consume. As the bottom line is enhanced, however, workers experience a problematic disconnect, Hochschild argues, between the emotions they display to the outer world and the feelings they experience within. This disconnect has parallels to recent studies by my colleague Ann Kring of schizophrenics. Contrary to longstanding assumptions about schizophrenia and flat affect, schizophrenics have been shown to feel the emotions that you and I feel but not to express them in the face. Service industry jobs produce a form of schizophrenia: We may experience feelings of emptiness and quiet frustration, or a deep ennui, but we display to the world the smile of satisfaction.

How then can we provide a coherent analysis of a category of behaviors—smiles—that includes my McD smile as well as the loving smiles of old friends and parents and children? At first glance, the empirical literature on the smile yields similarly paradoxical findings: People have been shown to smile when winning, losing, watching a film of an amputation, eating sweets, facing adversaries, experiencing pain, feeling affection toward loved ones. The answer is provided by Paul Ekman, and it involves looking away from the lip corners to that wellspring of the soul—the eyes.

A vocabulary of smiles comes sharply into focus when we consider the activity of the happiness muscle, the orbicularis oculi. This muscle surrounds the eyes and when contracted leads to the raising of the cheek, the pouching of the lower eyelid, and the appearance of those dreaded crow's-feet—the most visible sign of happiness—which the Botox industry is trying to wipe out of the vocabulary of human expression. People may think they look prettier following Botox injections, but their partners will receive fewer clues to their joy, love, and devotion.

Ekman has called smiles that involve the activation of the zygomatic major muscle and the orbicularis oculi the Duchenne or D smile, in honor of the French neuroanatomist Guillaume Benjamin Amand Duchenne (1806–1875), who first discovered the visible traces of the activity of orbicularis oculi. Smiles that do not involve the activity of the happiness muscle, the orbicularis oculi, are sensibly known as non-Duchenne or non-D smiles. To try your hand at this subtle distinction between Duchenne and non-Duchenne smiles, see if you can detect which is which in the photographs below (answers provided on page 108).

Dozens of scientific studies speak to the importance of parsing the heterogeneous category of smiles according to the activity of the orbicularis oculi muscle. Duchenne smiles differ morphologically in many ways from the many other smiles that do not involve the action of the orbicularis oculi muscle. They tend to last between one and five seconds, and the lip corners tend to be raised to equal degrees on both sides of the face. Smiles missing the action of the orbicularis oculi and likely masking negative states can be on the face for very brief periods (250 milliseconds) or very long periods (a

Smile quiz: Examples of Duchenne and non-Duchenne smiles.

lifetime of polite smiling by oppressed airline stewardesses and fast-food servers). Non-D smiles are more likely to be asymmetrical in the intensity of muscle firing on the two sides of the face.

D smiles tend to be associated with activity in the left anterior portion of the frontal lobes, a region of the brain preferentially activated during positive emotional experiences. Non-D smiles, in contrast, are associated with activity in the right anterior portion of the brain—a region associated with the activation of negative emotion. When a ten-month-old is approached by his or her mother, the face lights up with the D smile; when a stranger approaches, the same infant greets the approaching adult with a wary non-D smile.

And importantly, several studies have found that Duchenne and non-Duchenne smiles, brief two- to three-second displays differing only in the activation of the orbicularis oculi muscle, map onto entirely different emotional experiences. For example, in a long-standing collaboration with my friend George Bonanno, a pioneer in the study of trauma (see chapter 7), we interviewed middle-aged adults six months after their deceased spouse had passed away. These individuals were asked to describe their relationship with their deceased spouse for six minutes. I spent a summer coding the occurrence of Duchenne and non-Duchenne smiles from video-tapes of these narratives. We then related measures of bereaved participants' D and non-D smiles to their reports of how much enjoyment, anger, distress, and fear they felt during the interview, which we gathered immediately after the participants had finished talking about their deceased spouse.

Portrayed in the table below are the correlations between how much participants showed these brief Duchenne and non-Duchenne smiles and their ensuing self-reports of emotion gathered moments later. Positive scores indicate that the more they showed the particular smile during the six-minute interview, the more they subsequently felt the particular emotion listed on the left. Negative correlation values reveal the opposite, that the more the participant smiled in Duchenne or non-Duchenne fashion, the less of the emotion they felt. Asterisks indicate that the observed correlation was statistically significant, and not likely produced by chance.

	DUCHENNE SMILES	NON-DUCHENNE SMILES
ENJOYMENT	.35*	−.25*
ANGER	−.28*	.09
DISTRESS	−.49*	−.16
FEAR	−.31*	.04

What is impressive about these data is that very brief Duchenne smiles involving the activity of the orbicularis oculi were associated with increased feelings of enjoyment during the conversation, and reduced feelings of anger, distress, and fear. Non-Duchenne smiles were associated with the opposite pattern of experience—reduced feelings of enjoyment and none of the negative emotions.

The Duchenne/non-Duchenne distinction is the first big distinction in a taxonomy of different smiles. One kind of smile involves the orbicularis oculi muscle, and accompanies high spirits and goodwill. As we shall see, when other movements are added to the D smile, people can communicate different positive states like love, awe, and desire. A second kind of smile is the non-D smile, which reflects the attempt to mask some underlying negative state. In *Emotions Revealed,* Ekman deconstructed the non-D smile into a dizzying array of smiles, including pained smiles, fearful smiles, contemptuous smiles, and submissive smiles.

Twenty-five summers ago, as I served that reliable customer his four burgers and coffee, I am absolutely confident that not a trace of orbicularis oculi activity was to be seen on my late-pubescent face. I would have been an easy case study for Ekman; it would have been simple for him to reveal which negative states—despair, frustration, contempt—I was attempting to hide with my halfhearted McD smile. Off work, and at last with friends jumping off rocks into alpine rivers, I am sure the D smiles would have washed over my face. Studies inspired by Ekman's analysis would reveal that these D smiles are a glue of social life, and a provenance of the camaraderie that make me nostalgic for those carefree times.

THE SMILE AS SOCIAL CHOCOLATE

In the 1980s developmental psychologists Ed Tronick, Jeff Cohn, and Tiffany Field became interested in what postpartum depression does to mother-child interactions. Their studies, and those of other investigators, revealed that postpartum depression mutes the positive emotionality of the mother—she smiles less, she vocalizes with less warm intonation, and her positive emotional repertoire is less contingent upon the actions of her child. Children of mothers experiencing postpartum depression tend to show complementary behavior—they are more agitated, distressed, and anxious.

Answers to the smile quiz on page 106: For the first gentleman, the D smile is on the right; for the second, it's on the left.

In my view, the Dutch artist Jan Steen (1626–1679) is the greatest painter of human smiling and laughter. In *Merry Company on a Terrace* (1673–1675), which hangs in the New York Metropolitan Museum of Art, one finds sharply portrayed varieties of smiling and laughter. The maiden in the foreground, based on Steen's second wife (his first died early in life), shows a coy smile: the head is turned away, but there is eye contact. The jester in the upper left corner (with sausage pinned to his comical hat) smiles with a lascivious tongue protrusion, and close inspection reveals a slight pucker in lips of the nearby woman—a sign of her potential affection. The slight smile of the lute player involves, it would seem, a subtle pattern of gaze upwards and raised eyebrows, which we have found to signal rapturous states. And the expression of the gentleman in the white hat to the far left—a self-portrait of Steen—reveals the many delights of laughter, which we will cover in our next chapter.

This kind of result is compellingly intuitive. Any parent or friend who has been up close to this phenomenon, who has been in the living room of a depressed mother whose positive emotion is dampened and disengaged from that of her child, readily knows how

essential the exchanges of smiles, coos, touch, play faces, and interested and encouraging eyebrow flashes are to the parent-child
dynamic. Yet from a scientific standpoint, the finding—the mother's
impoverished positive emotional repertoire brings about anxiety
and agitation in the child—is plagued by alternative explanations.
Perhaps agitated, fussy infants produce muted positive emotionality and depression in the mother. Perhaps they both share some
genetically based tendency that predisposes their parent-child
interactions to lack mutual smiles, coos, touches, and play. Perhaps
their shared emotional condition is the product of deeper structural
causes—underpaid work, poverty, alienated or abusive husbands,
and the like.

So to study the role of smiling and muted positive emotionality
in parent-child interactions, Tronick, Cohn, and Field developed
what has come to be known as the still-face paradigm. This experimental technique is profoundly simple but powerful: The mother
is requested to simply be in the presence of her young infant, say
nine months old, but to show no facial expressions whatsoever, and
none of the most common of facial expressions for young mothers—smiles. As the young child navigates around the laboratory
environment, approaching toy robots and stuffed elephants and
brightly colored objects that make farm animal noises, the child
looks to the mother's face for signals about the environment. The
child seeks information in facial muscle movements about what is
safe, fun, and worthy of curious exploration, and what is not, and
the mother sits there impassionate, stone-faced, and unresponsive.

The results are astonishing. In a smile-impoverished environment, the young child no longer explores the environment, no
longer approaches novel toys or play structures; her imagination
shuts down. The child quickly becomes agitated and distressed,
often wildly so, arching his or her back and crying out. The child will
often move to the mother and try to provoke her, stir her out of her
stupor, with a vocalization or touch or encouraging smile. And as the
child begins to resign herself to the unexpressive condition of the
mother, she moves away from the mother, refusing eye contact, and
eventually falls into listlessness and torpor.

The same is true, albeit on a much smaller scale, with adults.

Friends of depressives find their interactions, research shows, to be unrewarding, and at times difficult to sustain. In conversations with individuals who show little positive emotion in the face or voice, participants engage in less responsive social behavior—playful laughs, smiles, head nods, knowing mutual gazes—and experience the conversations as unrewarding.

The smiles, and I should say D smiles, which punctuate our daily interactions—between parents and children, flirting strangers, friends sharing a silent moment of satirical commentary upon an acquaintance—are like social chocolate. With chocolate waiting to be enjoyed, young children, and a good many adults, will do just about anything with verve—mow through that side of vegetables, clean hamster cages, listen to long-winded adult stories, finish an odious task at work. The same is true of smiles; they are the first incentives toward which young children move, and that parents hungrily seek. In relevant research, when one-year-old infants sit at the edge of a visual cliff, a glass surface over a precipitous drop, with their mother on the other side, the infant immediately looks to the mother for information about this ambiguous scene, which looks both dangerous and passable. If the mother shows fear, not a single child will crawl across the glass surface. If the mother smiles, my Berkeley colleague Joe Campos finds, approximately 80 percent of the infants will eagerly cross the surface, risking potential harm, to be in the warm, reassuring midst of their mother's smile.

From the standpoint of the person smiling, we know from elegant work by Barbara Fredrickson and Robert Levenson that when people emit D smiles when experiencing stress, their level of cardiovascular arousal quickly moves to a more quiescent baseline. My hunch is that, as Darwin observed, with the D smile the individual exhales strongly, which calms stress-related physiology. We have also already seen that in the midst of a D smile, the smiler's left portion of the frontal lobes—a region of the brain that processes information about rewards and enables goal-directed action—is activated.

Perhaps more dramatic is what the smile does to the person perceiving it. The definitive work on this topic has been done by Ulf Dimberg and Arne Öhman, working in labs in Norway and Sweden, respectively. These investigators have pioneered techniques for presenting images of facial expressions to perceivers at incredibly fast speeds, outside of the perceivers' conscious awareness. Most typically, in what is known as the backward masking paradigm, they present slides of facial expressions (for example, facial displays of anger or the smile) for exceptionally brief periods—say, 100 milliseconds. Immediately following the image of the smile, another image is presented—say, of a neutral face, or a chair—which wipes out the participant's ability to consciously represent the image of the smile. In the backward masking paradigm, participants cannot tell you with any reliability what first image they have just seen; the smile (or comparison facial display) has only been perceived at the unconscious level. And yet people who have viewed smiles in this fashion are more likely to smile and report greater contentment and well-being, and in some studies, they show calmer cardiovascular physiology. They have no idea what they have just seen, but the smile has enhanced their well-being.

It goes deeper than this. Richard Depue and Jeannine Morrone-Strupinsky suggest that perceiving smiles in others, most likely of the Duchenne variety, triggers the release of the neurotransmitter dopamine, which facilitates friendly approach and affiliation. As one illustration, dopamine is activated in heterosexual males by viewing the smiles of attractive females. Smiles catapult individuals toward one another, and in the more intimate space produced by mutual smiles a more proximal set of behaviors—touch and soothing vocalizations—kick into action, soothing, calming, and triggering the release of opiates, which bring about powerful feelings of warmth, calmness, and intimacy.

When you see the exchange of D smiles between a father and a toddler on a swing, between two adults flirting in the corner of a room, between two friends laughing over their latest efforts at work or romance, or between two strangers navigating who goes through the door first or who takes the last egg roll at the buffet—one cannot help but be struck by the simplicity of social pleasure. Two smiles are exchanged within the span of a second or two, this small but universal element of decorum is honored, and the day continues. Within the bodies of those individuals, however, are reciprocally coordinated surges of dopamine and the opiates. Stress-related cardiovascular response reduces. A sense of trust and social well-being rises. The smile is the dessert of our social lives. It evolved as a neon-light signal of cooperativeness, it became embedded in social exchanges between individuals that give rise to closeness and affiliation. The right kind of smile is a common contributor to the numerator of the *jen* ratio and a gateway to the life well lived. And this was the hypothesis I tested in an unusual study of yearbook photos of graduates from a small women's college in 1960, just about to head out into a world that would turn tumultuous.

FLEETING MOMENTS OF THE COURSE OF LIFE

Ravenna Helson is a pioneer in the study of women's lives. In the early stages of her scientific career in the 1950s, she was interested in the intellectual creativity of women—an area almost entirely ignored by psychological science—and interviewed female pioneers in mathematics and the physical sciences. She then turned her scientific imagination to the question of how identity develops. Almost all of the longitudinal research on identity and the course of life had been done on men; the lifespan development of the other half of the species was a mystery. In 1959, a few years prior to publication of Betty Friedan's *The Feminine Mystique*, Ravenna initiated what would become the longest longitudinal study of the lives of women ever conducted, the Mills Longitudinal Study. This study has followed the lives of approximately 110 women who graduated from Mills College in 1959 and 1960 for the past fifty years, and continues to this day. It has led to basic discoveries about how identity shifts over the course of life for women, and how it remains the same.

In 1999 Ravenna stopped by my office with a generous offer. She noted, in her slight Texas drawl and gentle style, that she had gathered college yearbook photos of her subjects. She wondered whether I might be interested, with a student of hers, LeeAnne Harker, in exploring whether her Mills participants' smiles, captured when they were graduating from college, would say anything about the next thirty years of their lives. The more conventional side to my scientific mind predisposed me to politely decline. The premise that expressive behavior gathered in one instant in time (in the few milliseconds it takes for the shutter to open) in such an artificial context (having your yearbook photo taken by a stranger) could actually predict anything meaningful about an individual's life violated the most sacred laws of studying individuals scientifically. Within the study of individuals, it is canonical to sample a person's behavior many times and in a diverse and revealing array of contexts. A more representative sampling of observations guarantees more reliable inferences about who the person is. If you want to know whom to marry or what friend to travel with, you're best served by seeing them when grumpy in the morning after a bad night's sleep, when handling the stress of a conflict, when experiencing pain, around their mothers and ex-spouses, and when things go really well, and not just when sparkling in witty repartee at a cocktail party. Relying upon one yearbook photo as a potential measure of the person's identity was problematic in this regard, to say the least.

Also problematic was the notion of discerning muscle movements from static photos. All research on facial expression had relied upon video or moving pictures, in which the effects of the facial muscle movements are evident in the onset and offset of changes in the appearance of the face. In identifying D smiles, for example, one needs to see the crow's-feet, cheek raise, and lower eyelid pouch, all subtle judgments that are best made when one can see these actions appear and disappear in video over time.

Undaunted, LeeAnne Harker and I took a week to code the yearbook photos of 110 women, carefully looking for evidence of the activity of the zygomatic major muscle as well as the oribicularis oculi. This coding produced a score between 0 and 10 capturing the warmth of each woman's smile.

LeeAnne and I then took this measure of the warmth of the smile and related it to the treasure trove of measures Ravenna had gathered on these Mills alumnae when the women returned to the Berkeley lab, often flying in from great distances, when they were twenty-seven, forty-two, and fifty-two. This included measures of their daily stress, their personalities, the health of their marriages,

Above are two of the photos that we coded. The Mills alumna on the left received a score of 7, with a 3 for her zygomatic major muscle movement and a 4 for her orbicularis oculi. The woman on the right received a score of 4, with a 3 being given for her zygomatic major muscle action and a 1 for very slight activity of the orbicularis oculi.

and their sense of meaning and well-being as they moved into middle age.

What we discovered about the benefits of the warm smile would fit the analysis of the smile developed here, and would prompt readers of the study to rustle around their closets in search of their yearbooks. Warm D smiles promote high *jen* ratios and the meaningful life.

Mills alumnae who showed warmer, stronger D smiles when they were twenty reported less anxiety, fear, sadness, pain, and despair on a daily basis for the next thirty years. The smile mitigates anxiety and pain, most likely through the effects smiling has on stress-related cardiovascular arousal. The strong D smilers also reported feeling more connected to those around them; the smile helps trigger greater trust and intimacy with others.

The warmth of a woman's smile also predicted a rising trajectory in her sense that she was achieving her goals. Women with warmer smiles for the next thirty years became more organized, mentally focused, and achievement-oriented. Forget what people have told you about creativity and achievement emerging out of despair and anxiety. Not so. Dozens of scientific studies have found that people who are led to experience brief positive emotions are more creative, expansive, generative, synthetic, and loosened up in their thought. Our Mills women who showed warmer smiles reflected these benefits of positive emotion across their lives.

Our results concerning the relationships of the Mills alumnae were perhaps even more striking. These women were brought to UC Berkeley to spend a day with other individuals, as well as a group of scientists who wrote up personal narratives based on their impressions of the women. Women with warm smiles made much more favorable impressions upon the scientists in this context, suggesting that the smile enables more positive social encounters.

Turning to marriage, those women who displayed warmer smiles were more likely to be married by age twenty-seven, less likely to have remained single into middle adulthood, and more likely to have satisfying marriages thirty years later. Much has been made of the toxic effect on marriages of negative emotions like contempt and ceaseless carping and criticism. John Gottman and Robert Levenson can predict with 92 percent accuracy that a couple will divorce when the partners show high levels of contempt, criticism, defensiveness, and stonewalling. These negative emotions are like poison. What about marital happiness? Here Gottman and colleagues are starting to show that respect, kindness, and humor help married couples deal more effectively with conflict in their relationships. This was the story in our study: Women with warmer smiles had healthier marriages.

Finally, women with warmer smiles at age twenty reported a more fulfilling life at age fifty-two. Across young and middle adulthood, women prone to expressing positive emotions experience fewer psychological and physical difficulties and greater satisfaction with their lives.

You are probably thinking several things right now about this study. Most importantly, what did you look like in your yearbook photos? (I was wearing a velvet bow tie in one yearbook photo, with a silk disco shirt and a slightly uncomfortable smile.) More importantly, you should definitely be searching through that cardboard box of family memorabilia out in the garage to find out what your partner looked like, for that is likely to say a lot about your own current happiness.

What about the following alternative thesis—the just say yes thesis—that what we're observing in this pattern of results is simply women who say yes to everything, regardless of whether they truly endorse what they are saying, or are happy or not. Perhaps there is a certain group of our women who, in the desire to please others, smile, say they feel connected to others, report accomplishing their life's work, and report being happy in their marriages; but in actuality their lives are a neurotic mess of anxiety, self-deception, and despair. There is a measure of this tendency to say socially desirable things to others, and in fact, when we statistically controlled for the women's tendency toward this, all of the results held up. The warm smile has positive benefits independent of just being outwardly and inauthentically agreeable to others.

Okay, what about beauty? Physical attractiveness has been shown to have a host of benefits for individuals, from an increased number of friends to larger raises in the workplace. Perhaps it was the beautiful Mills grads who had the warm smiles, and thus, perhaps it was beauty, and not the warmth captured in the D smile, that produced the results that we observed. Perhaps the long-term benefits of the warm smile in this study simply reduced to being outwardly beautiful. As it turns out, beauty is remarkably easy to judge from photos. We had a group of undergraduates rate the beauty of the 110 Mills alumnae in our study. More beautiful Mills grads did indeed feel more connected to others, less anxiety, and greater well-being. Importantly, the warmth of a woman's smile still predicted less anxiety, increased warmth toward others, greater competence, and healthier marriages and increased personal happiness when we controlled for how beautiful the participants were. Warmth and kindness differ from physical beauty.

SMILES AND THE ORIGINS OF HAPPINESS

Sometimes the simplest questions are disarmingly hard to answer. A graduate advisor of mine once stopped me in my tracks with this one: "So you're studying emotion . . . answer this one. Why do orgasms feel good?" I mumbled something about the opioids, and dopamine, and oxytocin, and then collapsed into a state of blushing, bumbling confusion. He was asking about the origins of happiness and pleasure—where they come from, and what their basic elements are—and my answer wasn't much of an answer. Electrochemical signals in the brain and body cannot provide a satisfying answer about the nature of experience or, in this case, what the roots of pleasure and happiness might be. What are the deeper evolutionary contexts that led to the centrality of the smile in our social life? Where does happiness come from?

Darwin's genius was to describe the patterns of behavior we see today—patterns of affection, submissiveness, laughter, and smiling—and to trace those fleeting but precise, efficient, designed behaviors back in evolutionary time to their deeper roots, to the survival and reproduction-related contexts in which they arose. This kind of evolutionary analysis has revealed that the earliest primate smile is a submissive display subordinates use when nearing dominant primates, and fearing a jugular-threatening attack or the forceful backhand of a hairy arm. If this was the end of our search for the origins of the smile, we would be confronting the following conclusion: that the smile has its origins in the attempt to short-circuit threat, that the smile emerges out of a tremulous anxiety about being destroyed, that it is based in the most powerful strategy weaker individuals can resort to—submissiveness. Happiness, by implication, is simply the by-product of our attempts to navigate threats to our existence.

Let's call this thesis the Woody Allen hypothesis, thanks to his characterization of the intertwinement of suffering, happiness, and love so central to his brilliant movies, captured in the quote below:

To love is to suffer. To avoid suffering, one must not love. But then, one suffers from not loving. Therefore, to love is to suffer; not to love is to suffer; to suffer is to suffer. To be happy is to love. To be happy, then, is to suffer, but suffering makes one unhappy. Therefore, to be happy, one must love or love to suffer or suffer from too much happiness.

Of course, Woody Allen works this thesis to hilarious effect in his comedy. I'll line up with his most impassioned fans to see his latest comedy the first night of release to laugh at the absurdity of human happiness, love, and neurotic suffering. The Woody Allen hypothesis might seem solely the provenance of his comedic imagination, but in actuality this hypothesis—that anxiety and dread lie at the core of human happiness—is a long-standing assumption in the West about the elemental ingredients, the basic molecules, of happiness. This view holds that at the core of our experience of positive emotions are threat and anxiety; our positive emotions are layered on top of, emerge out of, are antidotes to, negative emotions like despair, fear, and anger.

For example, Silvan Tomkins, who helped forge the scientific study of emotion in the early 1960s, argued that positive emotions evident in smiles and laughs emerge with the cessation of negative states, such as anger and fear. As one example, laughter and our sense of amusement emerge out of the termination of anger. Someone angers you, your heart rate rises and your muscles tense, you're ready to throw a punch, and then in an instant, it's over, and you are suffused with a feeling of levity and amusement, the antimatter of anger.

One can push this reasoning back further, to earlier intellectual predecessors with farther reach. For Freud, many pleasurable experiences, flights of the imagination—the creation of fiction, unnervingly insightful dreams, or uplifting, perspective-altering jokes—as well as many acts of altruism are really mechanisms that fight off basic human anxieties about our inappropriate sexual urges, or unacceptable inclinations toward aggression and destruction. You write an uplifting piece of fiction or give left-over food to a panhandler: The motive driving those acts is the reduction of neurotic anxiety.

One can forgive Freud these notions in light of the Victorian culture that surrounded him and which was the fertile ground of his theorizing. One finds the Woody Allen thesis in more recent scientific inquiry. Terror management theory, a widely influential theory in social psychology, holds that many noble acts—intellectual creativity, philosophical and spiritual traditions, participation in old cultural forms like collective celebrations or our devotion to artistic and political groups—arise out of an anxiety about our inevitable demise, for these acts convince us of the possibility that we live beyond our own physical death. It is assumed in the study of parent-child attachment that the fundamental emotion that drives attachment processes between parents and children, friends and intimates, is anxiety. It is the dread of being abandoned to the perils of solitude that prompts infant smiles, coos, squeaks, and giggles, which bring parents near, and the touch and intimate, idiosyncratic nicknames and voices of romantic partners.

The Woody Allen hypothesis has deep roots in Judeo-Christian thought about original sin and the fall from grace. Within this framework, human nature is evil, sinful, and decaying. True happiness arises not in the present life but in the escape from the body and its corruptions. Happiness is found in a spiritual state freed from the sins of the flesh, in the afterlife—in communion with God. Happiness arises in the abandonment of the present moment, and when we are free of our earthly desires. In terms more friendly to psychological science: Happiness is to be found only in the quiescence of negative states like greed, anxiety, and anger.

As we conclude our search back in time for a precise understanding of the evolution of that most common of facial displays—the smile—we encounter a different view of the roots of happiness. We have one question left to answer: How did the first primate smile, the silent bared-teeth display, so intertwined with submissiveness, evolve into the Duchenne smile, our display of happiness? We return to Signe Preuschoft's subtle observations, which help illuminate how the smile was freed from anxiety and defense and became the display it is today.

Specifically, Preuschoft finds that in more hierarchical macaques, such as the rhesus macaque, there is a narrow use of the silent bared-teeth and relaxed open-mouth display. The silent bared-teeth display—the predecessor to our smile—is used only as an appeasement display. In these status-conscious monkeys, the smile is intertwined with anxiety and defense.

There are more egalitarian macaque species, however, such as the Tonkean macaque. In these macaques, hierarchies are flatter and power is equally distributed. This social condition more closely resembles the hierarchies observed in our hominid predecessors and contemporary hunter-gatherers—power differences are reduced, and equality is more pronounced. In egalitarian primates, food sharing is pervasive, alliances among subordinates are common, and social life consists more of negotiation than assertion of force. Preuschoft has found that in less stratified macaques, monkeys put the silent bared-teeth display to many new uses: to reassure, to affiliate, and to reconcile, as well as to appease. This is a standard evolutionary principle—that adaptations such as the silent bared-teeth display are put to new uses in a broader array of contexts to respond adaptively to shifting selection pressures. With the rise of primate equality, the silent bared-teeth display became freed from its one-to-one mapping to fear and submissiveness, and was extended into new social contexts that promote affectionate cooperation and affiliation. This display became a sign of friendly intent, and the trigger of behavioral processes that allow for close proximity and cooperation—grooming, embraces, hand clasping, and the like. In egalitarian primates, the silent bared-teeth display folded into affiliative, pleasurable exchanges.

The physical signature of human happiness is the D smile. The D smile did not originate in contexts that we today think are fast tracks to happiness. The D smile did not originate in experiences of sensory pleasure—Cro-Magnon individuals savoring fresh meat or the ripest of berries. It did not originate in contexts where our hominid predecessors enjoyed shifts upward in social status. The first D smile did not originate in contexts in which one individual enjoyed the accumulation of important resources. In fact, Christopher Boehm has summarized studies of hunter-gatherer hierarchies, and found that they systematically downplay any sudden abundance in resources through modesty and generosity.

In our primate evolution, the D smile was the first vocabulary of friendly intent and affection, in particular between near-equals. High *jen* ratios and the roots of human happiness are found in those moments when individuals moved toward one another toward cooperative and intimate ends. Our ultrasociality required this, as well as an all-purpose signal of cooperative intent, one that is highly visible and unambiguous, and one that could preempt conflict and spread cooperative relations potently and quickly, faster than a stranger could cock his arm and throw the first punch. Evolution's answer to the question of how to most powerfully communicate our capacity for *jen* was like that of the classical Greeks: the smile.

7

Laughter

IN THE 1982 FILM *Quest for Fire*, three hapless Neanderthal males leave their marsh-dwelling tribe in search of fire—the source of their haphazard provision of food and the hierarchical organization of their group. During their quest the three travelers escape from saber-toothed tigers, encounter towering woolly mammoths, and scare off a potential attack from a small tribe of paunchy, red-haired Neanderthals. In this last escapade, they rescue a different kind of early human. She is a more evolved female *Homo habilis*, finer in bone structure and facial morphology, lacking the carpet of hair covering the body, and adorned in patterned, tribal paints.

This female leads the three males on a primordial *Jules and Jim* road trip to her village. In this adventure, several distinctions between the Neanderthals and the *Homo habilis* come into sharp focus. The *Homo habilis* have developed special tools: a small board with a hole in it and a rounded stick to twist to create fire whenever needed—a radical innovation appreciated even by the dim-witted Neanderthals. They have more complex vocalizations than the grunts, groans, and growls of the Neanderthals. They beautify themselves with rudimentary paints. They live in sophisticated huts, organized in patterns comparable to that of the friendliest cul-de-sacs. They cultivate plants and animals—so critical, Jared Diamond argues, to shifts in the evolution of human culture. They prefer face-to-face sex. And they laugh.

In one scene, the three Neanderthals and their new consort are reclining in the dappled light of a shady tree, grooming, scanning the environment, picking bugs out of the air to eat. Out of the blue a rock bounces off one of the male's jutting foreheads, prompting a scratch on the head, a cursory look around, and then a return to a quiet state of mindless digestion. The *Homo habilis* witnesses this simplest form of humor (I spent a good part of my youth bouncing harmless objects—acorns, olives, Good & Plenties—off my brother's head), and breaks into laughter. The three Neanderthals have no idea what to make of the weird sounds emanating from her mouth.

The thesis that laughter represents a critical evolutionary shift in hominid evolution is not as far-fetched as one might imagine. It is a point that evolutionists Matthew Gervais and David Sloan Wilson have made. The laugh might rightfully lay claim to the status of tool-making, agriculture, the opposable thumb, self-representation, imitation, the domestication of animals, upright gait, and symbolic language—an evolutionary signature of a great shift in our social organization, accompanied by shifts in our nervous system. What separates mammals from reptiles are the raw materials of laughter—play, and the ability to communicate with the voice (when's the last time you heard the family gecko howl for a nibble of your salmon or purr for a scratch behind the ears?).

More striking is how human laughter differs from that of our primate relatives—gorillas, chimps, and bonobos. In the most rudimentary sense, the laughter of the great apes resembles our own. Their relaxed open-mouth displays and panting vocalizations look and sound quite familiar. They emit these displays in similar contexts as we do—when being tickled and during rough-and-tumble play. As with humans, chimps and apes are most likely to show open-mouthed play faces in developmental periods (adolescence) and times of day (leading up to feeding) where play can defuse conflict. Yet the laughter of chimps and apes is more tightly linked to inhalation and exhalation patterns than that of humans. As a result, it is emitted as short, repetitive, single-breath pants, and has little acoustic variety.

Human laughter, by contrast, is stunning in its diversity and complexity. It is a language unto its own. There are derisive laughs, flirtatious laughs, singsongy laughs, embarrassed groans, piercing laughs, laughs of tension, silent, head-lightening laughs of euphoria, barrel-chested laughs of strength, laughs that signal the absurdity of the shortness of life and the extent to which we care about our existence, contemptuous laughs that signal privilege and class, and laughs that are little more than grunts or growls. It is because of this heterogeneity that laughter has escaped simple theoretical formulation. It is the analysis of this heterogeneity that will lead to an answer about why we laugh.

LAUGHTER FACTS

In T. C. Boyle's *Drop City*, a community of hippies, devoted to free love, spontaneous ritual, and immersion in nature, moves from their compound, Drop City in Sonoma County, California, to the last outpost of unspoiled nature—arctic Alaska. This journey, an expression of the American spirit, provides ample opportunity for laughter amid the inevitable conflicts of free love and broken-down cars and all too earthly negotiations of who does the dishes in a commune devoted to passion and ecstasy. Boyle's descriptions of laughter reveal several insights about laughter:

> He heard Star laugh though, a hard harsh dart of a laugh that stuck right in him as he went off into the night, looking for something else altogether.

> Her first response was a laugh, musical and ringing, a laugh that made the place swell till it was like a concert hall.

> And then he began to chuckle, a low soft breathless push of air that might have been the first two bars of a song.

> There was a smattering of nervous laughter when he descended the steps and the laughter boiled up into a wild

irrepressible storm of hoots and catcalls and whinnying shrieks as the door pulled shut and Norm put the bus in gear and headed off toward the lights of Canada.

Star let out a laugh in response to something Jimmy had said, and then they were all laughing—even him, even Marco, though he had no idea what he was laughing about or for or whether laughing was the appropriate response to the situation.

A new round of laughter. Dale Murray joined in too, whinnying along with the rest of them.

Suddenly he let out a laugh—a high sharp bark of a laugh that startled the dog out of his digestive trance—and he raised his head and gave Marco a sidelong look.

"Big spender," she said, and her laugh trailed out over the river, hit the bank and came rebounding back again.

He then heard a squeal from Merry, or maybe it was Lydia, and a long sustained jag of laughter from all three of them, as if the very fact of his existence was the funniest thing in the world.

There were a few sniggers, a nervous laugh or two.

But they ate caribou tongue and Eskimo ice cream (caribou fat whipped into a confection with half a ton of sugar and a scattering of sour berries; Pan tasted it—"Ice cream, brother, it's ice cream," Joe Bosky told him, egging him on, but he spat it right back out into the palm of his hand, and the whole room went down in flames, laughing their asses off, funniest thing in the world, white man).

Pamela took one look at her and burst out laughing—she had to set down her cup because she was laughing so hard, her eyes squeezed down to semi-circular slits, her hands gone to her temples as if to keep her head anchored on her shoulders.

A first and perhaps most basic laughter fact is that nearly all laughter—darts, barks, sniggers, whinnies, hoots, jags, shrieks, catcalls—is social. Estimates indicate that laughter is thirty times more likely to occur around others than in isolation. We must move outside the individual's mind to understand ways in which laughter binds people together.

Laughter is contagious. Laughter spreads to others, it washes over them, it sticks in people like darts, it fills rooms with a certain quality, it prompts others to begin laughing for no reason intelligible to the conscious mind. In *Drop City*, laughter routinely boils up into rounds, cascades, and storms. Rooms swell with laughter like music halls.

Laughter produces a remarkable physical state. People laugh their heads and asses off. During laughter, the body goes limp. The

Jan Steen's portrayal of himself as a lute player wonderfully captures the relaxed state of physical collapse associated with laughter. It is fitting that Steen is such a talented painter of laughter, for he suffered many tragedies, including the early death of his first wife.

individual is incapable of any sort of motion. I've asked my daughters in the midst of a bout of being tickled to try to willfully carry out certain basic movements—whistle, wink, stick their tongue out at me—and they didn't come close. In the paroxysm of laughter, the body falls into a quiescent, otherworldly state.

And perhaps most subtly, laughter is intertwined with our breathing. In Boyle's descriptions, laughter accompanies pushes of air out of the mouth. With the exception of certain pathological laughs (Merv Griffin, Arnold Horshack on the 1970s sitcom *Welcome Back, Kotter*), almost all laughter occurs as people exhale. This simple laughter fact may seem incidental to our understanding of laughter, but in fact it is fundamental. Here's why.

Respiration and heart rate are two of the body's most essential rhythms. These two rhythms play off each other like the voices of singers in an a cappella group. When you breathe in, your heart rate rises. When you breathe out, your heart rate drops, as does your blood pressure, and you move toward a state of relaxation.

This lung-heart dynamic has made its way into book titles (Waiting to Exhale), aphorisms ("Take a deep breath"), ethical mottoes in grammar-school classrooms ("Take a breath and count to ten"), the advice coaches give to their players attempting the game-winning free throw (they systematically exhale), and the thousand-year-old breathing exercises of yoga practices. Exhalation reduces fight/flight physiology, especially heart rate, calming the body down. In fact, a series of studies in the 1970s and 1980s found that simply having individuals engage in deep breathing led to reduced blood pressure, stress, and anxiety, and increased calm.

When Robert Provine examined spectograms of different laughs—that is, their acoustic signatures—he took out the staccato bursts that we hear as "ha, ha, ha" or "tee, hee, hee." These on average last about .75 seconds. In any typical laughter "bout," there are three to four of these "calls." What Provine found underlying those bursts was a deep sigh. Laughter is the primordial breathing technique, the first "take a deep breath" exhalation. When chimps and bonobos show the open-mouth play face, they are altering their fight/flight physiology to reduce the chances of aggression and opening up opportunities for play and affiliation.

GRUNTS, SNORTS, AND A SPACE OF ITS OWN

We have encountered some basic laughter facts—it is almost always social, it collapses the body into a state of relaxation, it is intertwined with breathing. We still, however, have not answered the simplest of questions: What is the meaning of a laugh? What unites the remarkable varieties of human laughter? Clues to understanding a category of expressive behavior—be it a sigh, a tongue protrusion, the eyebrow flash, or the blush—emerge when scientists seek principles that unite the varieties of behaviors within that category. We can thank Jo-Anne Bachorowski for this kind of painstaking work on the complex acoustics of human laughter.

As air moves through the human vocal apparatus (see figure below), upon being pushed out by muscle contractions surrounding the lungs it is given a vibratory pattern through movements of the vocal folds. The speed with which the vocal folds vibrate gives the sound its pitch. These sounds are then given additional acoustic

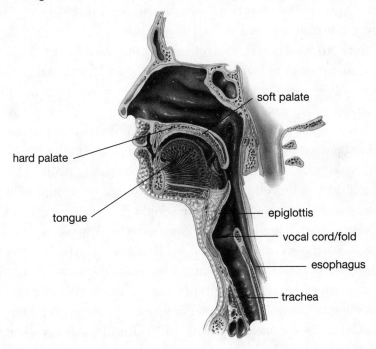

Diagram of the human vocal apparatus.

qualities, known as resonances and articulations, as they pass through the throat, delicate gymnastics of the human tongue, the opening of the mouth (for example, is it wide open, or are the teeth clenched?) and degree of opening in the nasal passage. Researchers then take these complex sounds, as represented in spectrograms, and extract a variety of different measures to arrive at an acoustic profile of a laugh, a sigh, a moan, a groan, or a tease. Measures include speech rate, pitch, loudness, pitch variability, and whether the sound rises or falls at the end.

Bachorowski was the first to put laughs through this complex form of acoustic analysis. She did so by recording the laughs of friends and strangers while watching Robin Williams, while playing amusing games together, or while simply talking casually. She has ruined her eyes in close-up analysis of thousands of laughs, and arrived at the beginnings of a laughter dictionary. There are cackles, hisses, breathy pants, snorts, grunts, and songlike laughs with mellifluous acoustic structure. Provine has found that women tend to laugh more than men, and Bachorowski's work ups the gender ante: Men, pitiful apes that they are, are much more likely to snort and grunt than women.

Bachorowski then conducted microscopic analyses of the fundamental acoustics of laughs. This laborious work yielded three clues to the deep meaning of laughter, and why it emerged in human evolution. The first clue helps us to begin to make sense of the astounding varieties of laughter. Bachorowski has differentiated between what she calls voiced laughs, which have tone to them and involve vibrations of the vocal folds (chords), and unvoiced laughs, which do not. Voiced laughs sound like songs, rising and falling as they move through space. Other people perceive these laughs as invitations to friendship and camaraderie. Unvoiced laughs—hisses, snorts, grunts—are not perceived as such. Much as the language of smiles is divided into Duchenne and non-Duchenne smiles, there are voiced laughs of pleasure and unvoiced laughs not involving pleasure. In his remarkable meditation on laughter, *The Book of Laughter and Forgetting*, Milan Kundera writes about two kinds of laughs. The laugh of the devil denies the rational order of the world. The laugh of the angel affirms the beauty of things and brings lovers, friends, and comrades together in common purpose in an elevated state above the earthly ground. Voiced laughs are Kundera's laughs of angels and unvoiced laughs those of the devil. Both are vital to the social contract.

Bachorowski made a second crucial discovery in analyses of how the laughs of individuals play off one another like the sounds of different instruments in an orchestra. The laughs of friends, as opposed to those of strangers, start out as separate vocalizations but quickly shift to become overlapping, intertwined sounds whose acoustic qualities mimic each other. Bachorowski deemed these laughter duets antiphonal laughter. This is the kind of laughter that unites people in affection. Friends, when responding to humor and levity, quickly find a common place in acoustic space for sharing laughter; their minds are united in two- to three-second periods of antiphonal laughter.

Finally, Bachorowski identified where laughs fall in acoustic space compared to consonants and vowels. Here a remarkable discovery: Laughs occupy a part of acoustic space that is different from vowel sounds like "ahhh" and "eee." We may describe laughs in the written word as "ha, ha, ha" or "hee, hee, hee," but in fact the acoustic structure of laughter is distinct from that of the vowels we use to represent this mysterious category of behavior. Certain regions of the human vocal apparatus produce the vowels and consonants that make up human speech, in which so much of human social life transpires. But there is another register of the human vocal apparatus, another form of output—laughter—with different origins and functions than human speech.

In light of Bachorowski's discoveries, it is now assumed that laughter preceded language in human evolution, emerging in early protohuman form some four million years ago. This is significantly earlier than when humans started to put together vowels and consonants into phonemes, and those phonemes into words and sentences. Recent neuroscientific data on laughter, summarized by Willibald Ruch, one of the leading laughter scientists, yields a similar conclusion about the early appearance of human laughter in evolution. Ruch has synthesized numerous brain studies of laughter. Some focused on the brain correlates of pathological laughter.

For example, people who suffer from a syndrome known as pseudobulbar affect will abruptly break into uncontrollable laughter in response to inappropriate stimuli—the tilt of a head, the movement of a hand, a trivial comment in a conversation. In other studies, laughter was observed following electrical stimulation to specific regions of the brain. When people laugh, subcortical, limbic regions of the brain and brain stem—most notably a region known as the pons, which is involved in sleep and breathing—are activated. These regions are much older evolutionarily than the cortical regions involved in language, suggesting that the deeper meaning of laughter is intertwined with breathing.

WHAT'S SO FUNNY ABOUT LAUGHTER

Laughter, then, is social and contagious. It empties the air deep in the cavities of our lungs, allowing heart rate and blood pressure to drop, the muscles of fight/flight exertion to go limp, and our psyche to fall into a calm state. These laughter facts fit nicely with the most enduring notion about the meaning of laughter, that it is the behavioral output of the experience of humor. Humor is as difficult to deconstruct as laughter, but there is consensus about the canonical structure of humorous acts: they involve some juxtaposition of contradictory propositions that produces a state of tension and ambiguity. The resolution of that contradiction then arrives in the form of a conceptual insight or punchline, the contradiction is resolved, and we laugh.

This hypothesis, that laughter serves to reduce tension, ran into some uncharitable data gathered by Robert Provine. Rather than restrict himself to the sterile confines of the laboratory, or rely on abstract, armchair conceptual analysis, Provine turned his astute ear to the laughter that occurs in the real world. He had three undergraduate assistants surreptitiously record bursts of laughter in malls, in friendly conversations on street corners, in the cafeteria banter of college students. This small band of laugh collectors recorded over 1,200 laughs in all. Provine transcribed these episodes into laughter narratives and then dissected what people were talking about just prior to laughter.

Humor often did precede laughter. Who wouldn't have laughed or at least chuckled with head tilted back, closed eyes, and collapsed torso and shoulders after hearing the following statements?

She's working on a PhD in horizontal folk dancing.

You just farted!

Poor boy looks just like his father.

When they asked John, he said that he wanted to grow up to be a bird.

Do you date within your species?

Was that before or after I took off my clothes?

Is that considered clothing or shelter?

Humor-oriented utterances, however, represented only 10 to 20 percent of prelaugh statements. Importantly, Provine found that laughter followed all sorts of utterances. Over 80 percent of the laughs did not occur in response to humor. Consider some of the following utterances that produced laughter.

I see your point.

I hope we all do well.

We can handle this.

I told you so!

Are you sure?

Why are you telling me this?

What is that supposed to mean?!

Not exactly knee-slapping fare ready for *The Daily Show, Saturday Night Live,* Robin Williams, the class clown, or the town wit. If these elicitors of laughter were just exceptions to the rule, we could readily discount them. But conversational events unrelated to humor were the rule and not the exception, and beg for more precise theorizing about laughter.

THE COOPERATION SWITCH

What, then, is the conceptual unifier of the cackles, guffaws, hisses, chortles, snorts, and melodious songs that we hear every day, or at least on the days that are more pleasing to the soul? We have seen that the time-honored thesis—humor—does not suffice. It fails to explain many, even the vast majority, of laughs that occur in our daily living. For Bachorowski and her colleague Michael Owren, the answer is cooperation. In an insightful analysis, Bachorowski and Owren argue that laughter builds cooperative bonds vital to group living. It does so through two mechanisms.

The first is contagion: We routinely laugh, and experience exhilaration and levity, at the sound of another's laugh. The contagious power of laughter motivated the introduction of laugh tracks on TV, a history that Provine details in *Laughter*. Recent neuroscience evidence suggests that when we hear others laugh, mirror neurons represent that expressive behavior and quickly activate action tendencies and experiences that simulate the original laugh in the listener's brain. Specifically, laughter triggers activation in a region of the motor cortex in the listener, the supplementary motor area (SMA). Bundles of neurons leaving the SMA go to the insula and the amygdala, thus triggering the experience of mirth and amusement in the perceiver of the laugh. When we hear others laugh, this system of mirror neurons acts as if the listener is laughing.

Laughter builds cooperative bonds through a second mechanism, Bachorowski and Owren propose: Laughter rewards mutually beneficial exchanges—successful collaborations at work, in the kitchen, in child rearing, with friends. Laughter signals appreciation and shared understanding. Laughter evokes pleasure. Given that each individual has a signature laugh, produced by the particulars of the vocal apparatus, laughs become unique rewards of cooperative exchange, building trust between individuals.

This theorizing yields deep insights into laughter. Laughter is not simply a read-out of an internal state in the body or mind, be it the cessation of anxiety and distress or uplifting rises in mirth, levity, or exhilaration. Instead, laughter is also a rich social signal that has evolved within play interactions—tickling, roughhousing, banter—to evoke cooperative responses in others. The laughter as cooperation thesis brings together scattered findings in the empirical literature. A deadlocked negotiation between Palestinian and Israeli negotiators took a dramatic turn toward common ground and compromise after they had laughed together. In my own research with executives, laughter early in negotiations—the product of breaking-the-ice banter about families, travel mishaps, hotel rooms, golf games, and the like—sets the stage for mutually beneficial bargaining. Workplace studies find that coworkers often laugh when negotiating potential conflicts—in tight spaces, at tense team meetings, when critiquing a colleague's work. Romantic partners who manage to laugh while discussing an issue of conflict find greater satisfaction in their intimate relations. Strangers who laugh while flirting in casual conversation report greater attraction. Friends whose laughs join in antiphonal form discover greater intimacy and closeness.

And what applies to the role of laughter in Middle East negotiations and the pyrotechnics of executives haggling, colleagues coexisting, and strangers flirting speaks to the long-term trajectories of attempts at connubial peace. John Gottman has found that for couples who were divorced on average 7.4 years after they were married, negative affect—for example, contempt and anger—was especially predictive of marital demise. For couples who divorced on average 13.9 years after they were married, it was the absence of laughter that predicted the end of their bond. In the early stages of a marriage, anger and contempt are highly toxic. In the later phases of intimate relations, it is the dearth of laughter that leads individuals to part ways. Without that cooperative frame for an intimate bond that laughter provides, as well as its attendant delights, partners move on.

Perhaps laughter is the great switch of cooperation. It is a framing device, shifting social interactions to collaborative exchanges based on trust, cooperation, and goodwill. Perhaps the pulse of a marriage is to be heard in the laughter the partners share. When I awaken and I hear my two daughters giggling in the antiphonal laughter that Bachorowski discovered, I know the morning will be fine, and relatively free of the conflicts of siblings as they seek their distinct niches in life. Perhaps our relationships are only as good as our histories of laughter together.

This theorizing, though, is in need of a bit more precision. We cooperate in many ways—through gifts, soothing touch, compliments, promises, and acts of generosity. Laughter must be associated with a more specific brand of cooperation.

Counterexamples to the laughter as cooperation hypothesis readily leap to mind. Bullies routinely laugh at their aggressive acts of humiliation (just listen to the shrill nerve jangling "ha, haa" of Nelson, the bully on *The Simpsons*). Some torturers at Abu Ghraib were heard to laugh at their victims. Thomas Hobbes wrote that laughter is the "sudden glory" produced by "the apprehension of some deformed thing in another" that makes people "suddenly applaud themselves"—a view that does not surprise given his portrayal of a dog-eat-dog world. Clues to a more precise conceptualization of laughter are found in its origins—in how play and laughter emerge in children, and what is being achieved, socially and conceptually, in the process.

THE ABUSE OF LANGUAGE

The acquisition of language in young children is breathtaking. Children learn ten or so words a day until the age of six, when the average child has a command of over 13,000 words. Children produce grammatically complex phrases even when not given such input from their parents, for example when parents speak pidgin. It is for these reasons that Steve Pinker called this high-wattage capacity the language instinct.

Just as remarkable, though, is how quickly children begin to abuse the rules of language. In particular, there is striking developmental regularity in the tendency for children, early in life, to violate basic rules of representation. They quickly start producing utterances that violate notions that words are supposed to refer to specific objects, and objects are to be characterized by specific words. And it is in this representational abuse that we find the core meaning of laughter—laughter indicates that alternatives to reality are possible, it is an invitation to enter into the world of pretense, it is a suspension of the demands of literal meaning and more formal social exchange. Laughter is a ticket to travel to the landscape of the human imagination.

In his analysis of the development of pretense, Alan Leslie details three kinds of pretend play in children. Each kind of pretend play hinges on the child violating the rules regarding correspondences between words and the objects to which they refer. In object substitution, the child substitutes nonliteral meanings of objects for the real meaning of the object. In the young child's world of pretend play, rocks become bread, swim goggles become cell phones, pillows become walls to fortresses, bedrooms become classrooms, older sisters petty rock stars or demanding old dames in the grocery store they run in the living room.

Children attribute nonliteral properties to objects in a second form of pretend play. While my daughters were five and three, respectively, I spent the better part of a year being a prince dancing with them at various balls. They insisted that I wear a certain pair of sweats, which they ascribed with the velvety beauty of a prince's medieval tights. This form of play, founded on the attribution of pretend properties, shifted a bit later to a set of identities I felt much more at home in—the ogre or friendly gorilla—all pretend identities that derived from elaborations upon my physical status and regrettable postpartum paunch.

And finally, the young child's world becomes filled with imaginary objects. In this third kind of pretend play, children simply imagine things that are not there—chalices in the princess's cupped hand, swords, magic carpets, evil witches and comrades in common cause.

These forms of pretense emerge in systematic fashion at around eighteen months of age. They are all systematically accompanied by laughter. And they lead the child to develop the ability to use words to refer to multiple objects. As children free themselves from one-to-one relations between words and objects, they learn that words have multiple meanings. They also learn that objects can be many things—a banana can be a banana, a phone, an ogre's nose, or a boy's penis (when the parents aren't around).

In the freedom of pretend play, children learn that there are multiple perspectives upon objects, actions, and identities. The child moves out of the egocentrism of his or her own mind and learns that the beliefs and representations of other minds most certainly differ from one's own. And it is laughter that transports children to this platform of understanding and epistemological insight.

Developmental psychologists who have studied the pretend play of siblings in the home, or the playful wrestling of parents and children, or the playful exchanges of children on the playground, find that laughter reliably initiates and frames play routines. A child or a parent will laugh as a chase game, roughhousing, round of silly wordplay or storytelling gets under way. Linguist Paul Drew carefully analyzed the unfolding of family teasing interactions and found that they are framed by laughs. Laughter is a portal to the world of pretense, play, and the imagination; it is an invitation to a nonliteral world where the truths of identities, objects, and relations are momentarily suspended, and alternatives are willingly entertained. Those hours of pretend play—peek-a-boo games, monsters and princesses, the ogre under the bridge, astronauts—are the gateway to empathy and the moral imagination.

LA PETITE VACATION

In the observation that laughter accompanies the child's capacity to pretend, to participate in alternatives to the realities referred to in sincere communication, we arrive at a hypothesis about laughter. Let's call this hypothesis the laughter as vacation hypothesis. The name of this hypothesis honors the comedian Milton Berle, witness, it is safe to claim, to millions of laughs during his career. Summing up the mysteries of laughter, Berle proposed, "Laughter is an instant vacation." If orgasm for the French is la petite mort (the little death), laughter is la petite vacation.

The wisdom of Berle's hypothesis is found in the etymology of "vacation," which yields a nuanced story. The word "vacation" traces its linguistic history back to the Latin *vacare*, which means to be "empty, free, or at leisure" and is defined as a formal suspension of activity or duty. The laugh, then, signals the suspension of formal, sincere meaning. It points to a layer of interaction where alternatives to assumed truths are possible, where identities are light-hearted and nonserious. When people laugh, they are taking a momentary vacation from the more sincere claims and implications of their actions.

So let's weave our facts and speculations together into the petite vacation hypothesis. In our primate evolution, laughter begins in the open-mouth play faces of chimps and bonobos, which signal and initiate playful routines. The quality of laughter, its sound and function and feeling, is rooted in physical action, as Darwin long ago observed: It is intertwined with exhalation, and the reduction in stress-related physiology. A special realm of sound is reserved for laughs, and it is an ancient one that predates language, represented in old regions of the nervous system—the brain stem—which also regulates breathing. This acoustic space reserved for laughs triggers laughter and pleasure in others, and designates, like the confines of a circus or theater, a social realm for acts of pretense and the imagination. In the pretend play of young children, laughter enables playful routines that allow them to have alternative perspectives on the world they are facing. Laughter is a ticket to the world of pretense, it is a two- to three-second vacation from the encumbrances, burdens, and gravity of the world of literal truths and sincere commitments.

LAUGHING AT DEATH

My dear friend and colleague George Bonanno took a while to get to academics. After riding trains, picking apples in Washington

State, living in communes, and painting signs in Arizona, he decided, on a whim, to take a community college creative writing course. After his first submission, he was discovered by his instructor, and quickly found himself on a fast track toward a PhD at Yale. Proponents of the conventional view of trauma may have wished that he never took that writing course.

For the past fifteen years, using intensive narrative interviews and longitudinal designs, he has studied how individuals adapt to various kinds of trauma—the death of a marital partner, the 9/11 attacks, sexual abuse, the death of a child. He kept encountering a basic finding not anticipated in the literature on trauma. The conventional view is that after a trauma everyone suffers prolonged periods of maladjustment, anxiety, distress, and depression. George has found in every study he has conducted that a significant proportion of people suffering a trauma experience distress and upset but, in the broader scheme of things, fare quite well. Within a year, they are as happy as they were, more poignant perhaps, filled with bursts of breathless longing, but in the end, content with life, and perhaps a bit wiser.

His question: What allows people to adjust to life-altering traumas? Our answer: Laughter. Laughter provides a brief vacation from the existential impossibilities, the deep sadness, the disorienting anxieties, of losing a loved one, or losing a city or way of life.

To test this thesis, George and I undertook a study to look at the role of laughter during bereavement. To do so, we brought forty-five adults to our laboratory, individuals who six months prior had watched their spouses die. Six months into bereavement is a poignant time. The death of a spouse leaves individuals mildly depressed, disoriented, lonely, and disorganized. The daily rhythms of a marriage are gone. So too are the conversations about what happened during the day, the fragments of a dream, the funny thing a friend or loved one did or said, how work went. Bereaved adults often have trouble conducting the daily affairs of their lives— remembering to pay bills, plan dinners, go shopping, fix cars— because the other part of their collective mind is gone. Reminders of their partner—photos, clothing, scents and sounds from the past—weigh them down in yearning. So we asked: Would laughter prompt bereaved adults to find new layers of meaning in the midst of trauma, and perhaps a path to the meaningful life?

Our forty-five participants came to George's lab in San Francisco, really an upstairs room in an old Victorian, with wood floors and paned glass. After some preliminary talk, George asked the participants the simplest of things, to "tell me about your relationship with your deceased partner." They were then given six minutes to tell their narratives of their relationships with the deceased spouse. There were stories of meeting one another at a blues show, of wild youth, raising children, and then bleeding gums that presaged a rapid death six months later, with children at the mother's side at the hospital bed. One man, in response to George's question, could only sob and gasp for six minutes, uttering not a word. I remember another woman whose husband had committed suicide at the end of a manic episode that was capped off by a disturbing visit to his mother. At the end of her narrative of this freefall, one could hear doves cooing on the windowsill of the lab room.

As George planned the next stages of this longitudinal study (he has assessed the well-being of these individuals for several years), he sent me the videotapes of these conversations. For an entire summer, locked up in my laboratory video coding room in the basement of my department, I coded these six-minute conversations with Ekman and Friesen's Facial Action Coding System. Each conversation took about six hours to code. Spending eight hours each day listening to stories of dying and coding such deep emotion left me exhausted and humbled. Almost all of our participants showed numerous displays of negative emotion, such as anger, sadness, fear, and, less frequently, disgust.

Our question was a simple one that had never been addressed before: What emotions predict healthy adjustment to the death of a spouse, as assessed with clinically sound measures of anxiety and depression, as well as measures of prolonged bereavement, which captures the individual's continuing longing for the deceased and inability to reenter into daily living? And which emotions predict poor adjustment during bereavement?

Traditional bereavement theories offer two clear predictions. This thinking is based on Freudian notions of "working through" the emotional pain of loss and the cathartic release of anger. It predicts that recovery from bereavement depends on the increased expression of negative emotions, such as anger and sadness. A second prediction is that the expression of positive emotion is in actuality a pathological sign of denial, of an intentional turning away from the existential facts of trauma, and impedes grief resolution. Our thinking was just the opposite, that laughter would allow our bereaved participants to distance themselves momentarily from the pain of the loss, to gain perspective, to look upon their lives in a more detached way, to find a moment of peace, to take a deep breath, so to speak.

Our first finding lent support to this view of laughter. Measures of laughter (and smiling) predicted *reduced* grief as assessed at six, fourteen, and twenty-five months postloss. Those participants who showed pleasurable, Duchenne laughter while talking about their deceased spouses were less anxious and depressed, and more engaged in their daily living, for the next two years. Just as important, people who showed more anger were observed to be experiencing more anxiety, depression, and disengagement from daily living for the next two years.

A first objection one might raise with respect to these findings concerns the nature of the death. Perhaps those individuals who laughed had partners who experienced easier deaths and thereby felt less initial grief and, as a result, were better able to adjust to this difficult loss. We know from empirical studies of bereavement that the nature of the death matters—sudden deaths, and deaths that produce greater financial demands upon the spouse, lead to prolonged grief and difficulty readjusting. We also know that the severity of the individual's initial grief powerfully predicts the degree of difficulty in adjusting that that person will experience later. These possibilities did not explain away the benefits of laughter: Those individuals who showed pleasurable laughter compared to those who did not did not differ in the nature of their spouse's death (its unexpectedness or financial impact) nor in their initial levels of grief.

One might likewise argue that perhaps our individuals who laughed at death were just happier individuals to begin with. Perhaps our results linking laughter to adjustment were simply the products of the temperamental happiness of the individual and not the emotional dynamics and perspective shifts accompanying laughter. This alternative too proved untenable—our people who laughed did not differ on any conventional measure of dispositional happiness from our individuals who did not laugh.

Buoyed by these findings, George and I went on a search for further evidence in support of the benefits of laughter. Why did laughing while talking about the deceased partner relate to increased personal adjustment? What we observed were findings very much in keeping with the laughter as vacation hypothesis. Our first analysis looked at how bereaved individuals' experience of distress tracked one physiological index of arousal—elevated heart rate. The bereaved individuals who laughed showed similar heart rate arousal as those who did not laugh. But whereas our nonlaughers' feelings of distress closely tracked increases in their heart rate, our laughers' feelings of distress were decoupled from this physiological index of stress. Metaphorically, laughers were taking a vacation from the stress of their partners' deaths, freed from the tension of stress-related physiology.

We then transcribed their conversations and identified exactly what the bereaved participants were talking about when they laughed. Here again, data suggest that laughter is not a sign of denial of trauma, as widely assumed, but an indicator of a shift toward a new perspective enabled by the imagination. We coded participants' references to several existential themes related to bereavement—loss, yearning, injustice, uncertainty. We also coded for insight words that reflect a shift in perspective, phrases like "I see" or "from this perspective" or "looking back." Our participants who laughed were most likely to be talking about the injustice of death—the unfair termination of life, the difficulties of raising a family alone, the loss of intimacy—but they engaged in this discourse with perspective-shifting clauses. Laughter was part of these individuals' shift in viewing the death of their spouses. It was a portal into a new understanding of their lives. A laugh is a lightning bolt of wisdom, a moment in which the individual steps back and gains a broader perspective upon their lives and the human condition.

Finally, our data speak to the social benefits of laughter. Our bereaved individuals who laughed reported better relations with a current significant other. They more readily engaged in new intimate relations.

LAUGHTER=NIRVANA

The Buddha's path to enlightenment was arduous. He had to leave the comforts of his well-to-do family, his wife, and his new child. He wandered for years grasping for the state of nirvana in different spiritual practices. He nearly died in ascetic practice, starving himself to bone on skin. When the Buddha finally attained enlightenment under the bodhi tree, it was in the realization that the suffering of life is rooted in self-centeredness and desire and that, once shed of such illusions, goodness arises from within. Loving kindness, compassion, right talk and action, peace, and indescribable joys are realized. In this epiphany the Buddha must have deeply exhaled. My bet is that he laughed as well.

Nibbana—nirvana—originally meant "to blow out." Clearly "blow out" refers to blowing out of the flames of self-interested desire, the obstacle to nirvana. I'd like to think a second possibility is that nirvana means to blow out, to exhale, to laugh.

Images of the Buddha are often images of full-bellied laughter. Study the images of the Dalai Lama with heads of state from around the world and they are all images of body-shifting laughter. The 100 Zen koans amassed in twelfth- and thirteenth-century Japan were used by Buddhist teachers to disengage the conscious rational mind, opening up opportunities for enlightenment. Well-known koans are intentionally paradoxical:

If you meet the Buddha, kill him.

Two hands clap and there is a sound. What is the sound of one hand?

Many other koans employ absurd humor; they have survived because of their capacity to reduce disciples to laughter:

What is the Buddha? Three pounds of flax.

What is the Buddha? Dried dung.

Laughter may just be the first step to nirvana. When people laugh, they are enjoying a vacation from the conflicts of social living. They are exhaling, blowing out, and their bodies are moving toward a peaceful state, incapable of fight or flight. People see their lives from a different point of view, with new perspective and detachment. Their laughter spreads to others in milliseconds, through the firing of networks of mirror neurons. In shared laughter people touch, they make eye contact, their breathing and muscle actions are in sync, they enjoy the realm of intimate play. Conflicts are softened, and often resolved. Hierarchies negotiated. Attraction and intimacy are created. What was once in the denominator of the *jen* ratio—conflicts, tensions, frustrations—fades away. People move closer to one another in peaceful ways.

8

Tease

MALE PEACOCKS are well known for their outlandishly beautiful tails—hypnotically patterned signs of their genetic fitness, alluring to the more dowdy and modest peahen. Less well known is how provocative the florid peacock male can be during the ritualized patterns and exchanges of peacock courtship. Often, when an inquiring peahen approaches, he will turn his back on her, as disinterested as the coldest of cold-shouldering high-school girls. He then stretches out his expansive tail to reveal to her inquisitive eyes his backside. What does the peahen get upon expressing initial interest? Very often, a nose brush with the male peacock's unseemly behind.

Why such a lack of bird decorum? Is the male peacock relying on the rump presentation as a sexual stimulant, as many primate species do, the most dramatic example being the baboon? Not so, reason Amotz and Avishag Zahavi, in their wonderful *The Handicap Principle*. These ornithologists suggest that the male peacock is simply testing the peahen. He is teasing and provoking her to gather information about her sexual interest. If the female, facing her consort's derrière, circles around to face the male with alacrity and earnest intent, the male knows that she is interested, and not just playing the field or stopping by for a casual exchange of clucks and coos. If she fails to appear, or does so after a few additional milliseconds of dillydallying, he has acquired critical information about her lack of commitment. He can factor this information into his decision about whom to mate with and whom not.

Lest you think that humans have evolved beyond the need to provoke and tease in intimate affairs, consider this exchange between two of literature's great lovers, Beatrice and Benedick from Shakespeare's *Much Ado About Nothing*. This revealing exchange is their first declaration of their affection for one another.

BENEDICK   And, I pray thee now tell me, for which of my bad parts didst thou first fall in love with me?

BEATRICE   For them all together, which maintain'd so politic a state of evil that they will not admit any good part to intermingle with them. But for which of my good parts did you first suffer love for me?

BENEDICK   Suffer love! A good epithet! I do suffer love indeed, for I love thee against my will.

BEATRICE   In spite of your heart, I think. Alas, poor heart, if you spite it for my sake, I will spite it for yours, for I will never love that which my friend hates.

BENEDICK   Thou and I are too wise to woo peaceably.

The importance of provocation and teasing in our social evolution is suggested by how pervasive teasing is in the animal world. Chimpanzees dangle their tails, tickling noses and eyes, to provoke response in slumbering or distracted chimps nearby. African hunting dogs and dwarf mongooses jump all over each other in piles of playful provocation prior to a hunt, much like pad-slapping football players moments before the kickoff, provoking readiness to attack and defend. In humans, mothers will pull their breast away from weaning babes as they pucker up for a drink. Adults will play hide the face, peekaboo games to stir a sulking child. Teenage girls and boys resort to hostile nicknames and outlandishly gendered imitations to assess the romantic leanings and sexual experiences of their friends. Sexual insults are as reliable an occurrence in human social life as food sharing, greeting gestures, patterns of comfort, flirtation, and the expression of gratitude.

Teasing has long occupied a problematic place in Western culture. In Roman times, law prohibited *mala carmina*—abusive songs and poems that centered upon ritualized insult. Today teasing is often prohibited on the grammar-school playground and in the workplace. It is regulated by speech codes on college campuses. Irony, a relative of teasing, is not enjoying the most sterling of reputations. In the circles of literary criticism, a widely read treatise, "Regulations for Literary Criticism in the 1990s," lists Regulation VII as "no irony." The accompanying rationale is that "great literature demands of us a high seriousness of purpose—not disrespectful laughter and clowning around." In *For Common Things*, Jedediah Purdy, fresh from his undergraduate days at Yale, issued a clarion call for sincerity and a move away from the derisive irony that fills the air during the swilling of cocktails at Ivy League parties.

The perils of teasing are patently clear. "Just teasing" is invoked as a last defense by the grammar-school bully and the incorrigible sexual malfeasant at work. But what they are referring to with the claim "I was just teasing" upon closer inspection is not teasing at all but aggression and coercion, pure and simple. Bullies steal, punch, kick, spit on, torment, and humiliate. They don't really tease. Sexual predators grope, leer, and make crude, at times threatening, passes. They're pretty ineffectual flirts. In contrast, teasing is a mode of play, no doubt with a sharp edge, in which we provoke others. We turn to the playful provocation of teasing to negotiate the ambiguities of social living—establishing hierarchies, testing commitments to social norms, uncovering potential romantic interest, negotiating conflicts over work and resources. To understand how this is so, we must first consider a universal institution that is the close relative of the tease—the jester or fool—as well as the philosophy of language. In doing so, we will discover a register of the voice and a pattern of semantics that illuminate the brilliant ways that humans put their bodies and representational minds to use in teasing.

FOOL'S PARADISE

On January 19, 1449, the Scots passed the Act for the Away-Putting of Feynet Fools. This act set into law punishments—the nailing of the ear to a post, amputation of fingers—for individuals falsely posing as jesters and fools. Fools and jesters were serious business in the Middle Ages and early Renaissance. Court jesters often acted as advisors on economic and diplomatic matters. They enjoyed well-remunerated positions within the courts of kings and queens in China, the Middle East, and Europe. The prominence of the jester and fool in public life is a human universal that dates back to the Aztecs, Mayans, and Native Americans of North America.

Court jesters, as richly detailed by Beatrice Otto in *Fools Are Everywhere*, hailed from poor backgrounds. Jesters most often were unusual in appearance or manner: hunchbacks, dwarves, and extremely ugly individuals were more likely candidates for this essential role than the town hunk (thus placing fools outside the competition for mates and resources). They often possessed other creative talents—they tended to be gifted musicians, poets, jugglers, or dancers.

Jesters dressed in their easily identifiable absurd garb. With their riddles, pratfalls, pranks, and sharp-tongued mockery, delivered with comical expressions, they satirized the powerful—the royal court, its nepotistic hangers-on, and especially the church. Jesters pointed to alternatives to the status quo. They reversed reality, turn-

An engraving of a court jester. Court jesters mocked the powers that be with their foolish garb, appearance, dancing, and spoken word.

ing conventional wisdom on its head. They often did so on behalf of the downtrodden and poor (and in fact, political pamphlets were developed out of some jesters' activities). In the words of the great court jester Nasrudin, "I'm upside down in this life."

When I began my study of teasing some ten years ago, the field would have benefited from the insights of a medieval jester or fool. They embodied a playfully provocative mode of commentary that speaks to the essence of what a tease is. The scientific study of teasing was hampered by poorly specified definitions of this ethereal phenomenon. This often happens when scientists rely upon natural language—the words we use—to capture what is largely a nonverbal phenomenon whose multilayered meaning is discerned in the subtle timing of a laugh or the shift in the speed or register of the voice.

The consensus was that teasing is "playful aggression." Clearly, though, teasing does not equate to all kinds of playful aggression. Unintended playful aggression—accidentally elbowing a fellow train passenger's nose while you're hustling money with your imitation of Harpo Marx—is clearly not teasing (at least I hope you don't think so). More general references to play are ambiguous. Many forms of childhood play, such as role playing (children acting as princesses or ninja warriors), roughhousing, highly structured playground games like tag or four square, and the ritualized jokes and conversational games that fill the air of school buses—are not teasing. The same is true of many forms of adult play: We tell amusing stories, exchange playful repartee, and josh around in ways that are not teasing.

Playful aggression fails to capture the edge of the antics of the jester and fool. In terms more felicitous to scientific inquiry, my colleagues Ann Kring and Lisa Capps and I defined a tease as an intentional provocation accompanied by playful off-record markers. We referred to provocation instead of aggression because a tease involves an act that is intended to provoke emotion, to discern another's commitments. The provocation is evident in the content of the verbal utterance or some physical act, like a poke in the ribs, the proverbial pinch of the cheek, or a tongue protrusion. The tease, in a funny way (and I'm not teasing), is like a social vaccine.

Vaccines are weak forms of pathogens (for example, smallpox) that, when injected, stimulate the recipient's immune system—the inflammation response, killer T cells that recognize the dangerous pathogen, bind to it and kill it. The tease seeks to stimulate the recipient's emotional system, to reveal the individual's social commitments.

The more mysterious element is what is unsaid in the tease. This family of linguistic acts we called off-record markers. These are the nonverbal actions that swirl around the hostile provocation and signal that it is not to be taken literally but instead in the spirit of play. Here we turned to philosophical accounts of literal and nonliteral communication to find principles that account for the art of the tease, philosophical principles that organize the antics of the fool and that help differentiate the bully from the sage.

THIS AND NOT THIS

In the mid-1960s, philosopher Paul Grice outlined four principles of communication that would profoundly shape the study of pragmatics—that is, how people speak. Sincere communication, according to Grice, involves utterances that are to be taken literally. These statements should adhere as closely as possible to four maxims (see table below). Statements should follow the rule of quality—they should be truthful, honest, and based in evidence. Statements should be appropriately informative—the rule of quantity—and avoid the Strunk and White catastrophes of being too wordy or opaquely succinct. Statements should be relevant and on topic and avoid meandering into digressions, irrelevances, or stream-of-consciousness flights of fancy. Finally, in honoring the rule of manner, statements should be direct, clear, and to the point (sorry if I've violated this one).

Utterances that follow these four simple rules are on-record, and are to be taken literally. When an MD provides a prognosis about a life-threatening condition, she should follow these four rules of on-record communication. So too should the financial advisor announcing the unexpected loss of a family fortune—

these are not the best moments for exaggeration, intentional false-hoods, fantastical description, obvious repetition, digression, meandering, or catchy metaphors or poetic obliqueness. Much of our social life, in fact—romantic declarations, sealing business deals, critiques at work, teaching young toddlers reaching to touch red-hot burners or rabid dogs—transpires in this realm of literal, on-record communication.

GRICE'S MAXIMS OF COMMUNICATION

LINGUISTIC PRINCIPLE	CRITERION	VIOLATIONS
QUALITY	TRUTH	EXAGGERATION, FANTASTICAL DESCRIPTION
QUANTITY	INFORMATIVENESS	REDUNDANCY, REPETITION, EXCESSIVE BREVITY
RELATION	RELEVANCE	DIGRESSION
MANNER	CLARITY	VAGUENESS, OBLIQUENESS, METAPHOR

When we intentionally violate Grice's maxims, we signal that alternative interpretations of the utterance are possible. We say "this" with our words, and "not this" with violations of Grice's maxims, pointing to other possible meanings of our utterance. We signal "not this" by resorting to obvious falsehoods or exaggerations of the truth (which violate the rule of quantity). We can provide too much information, for example in systematic repetition, or too little information, thus violating the rule of quantity. We can dwell in the irrelevant to violate the rule of relation. And we can resort to various linguistic acts—idiomatic expressions, metaphors, oblique references—that violate the rule of manner and its requirements of clarity and directness.

As important as sincere speech is to our social life, so too is this realm of nonliteral communication. Our brief utterances can take on the opposite meaning of what the words denote (irony, satire). We can connect disparate concepts in communicative acts that leap beyond narrow literal denotation (metaphor). We can endow our utterances with multiple layers of unbounded, aesthetically pleasing meaning (poetry).

The relevance of Grice's maxims to teasing, ironically enough, is revealed in linguists Brown and Levinson's 1987 classic, *Politeness.* Brown and Levinson carefully document how in the world's languages speakers add a layer of politeness to their utterances when what they say risks embarrassing the listener or themselves. Politeness is achieved through systematic violations of Grice's four maxims.

Consider the simple act of making a request. If someone asks you for the time, or directions, or to pass the rutabagas, or not to talk so loudly during the previews, that act is fraught with potential conflict. The recipient of the request is imposed upon and risks being revealed as incompetent, boorish, or disinterested in social conventions. The requester risks being perceived as intrusive and impolite. To soften the impact of requests and other potentially impolite acts such as recommendations, or criticism, people violate Grice's maxims to communicate in more polite fashion. Say your best friend is being a bit boisterous, with elbows flying, at your Friday evening line dancing group that you've generously invited him to. To encourage a bit more restraint, you might politely resort to indirect questions ("Have you ever seen yourself dance?"), rhetorical questions ("Have you done line dancing before?"), metaphors ("Wow, you holler like a howler monkey") and obliqueness ("I bet you'd be a terrific clown"). We break the rules of sincere communication to be polite. Equipped with this analysis of nonliteral communication, a careful examination of the tease reveals that teasing and politeness are surprisingly close relatives.

THE ART OF THE TEASE

What gives the tease the playful genius of the jester's satire are systematic violations of Grice's maxims. A first principle is exaggeration, which marks the playfulness of the tease by deviating from Grice's rule of quality. Teasing can involve copious detail, excessive profanity, or an exaggerated characterization. In a study of the conversations of a very loving family, the mother referred to a young son as "horse mouth" when he did not speak clearly. We tease with dramatic and exaggerated shifts in our pitch—we mock the plaintiveness of another with high-pitched imitations, and the momentary obtuseness of another with slow-moving, low-pitched utterances. Parents will tease children about their excessive possessiveness by using vowel elongation and exaggerated pitch: "Mine!" We tease by imitating, in exaggerated form, the mannerisms of others—the bread and butter of a preteen's around-the-clock, eye-rolling mockery of his or her parents.

Exaggeration is core to understanding "playing the dozens," a sophisticated form of ritualized insult that the sociologist Roger Abrahams documented while spending two years living among young black kids in urban Philadelphia in the early 1960s. Abrahams found that young black males, in particular between the ages of eight and fifteen, resorted to a canon of teases—"the dozens," oft-heard, profane poems about the target and the target's mother. These ritualized insults occurred only among friends, and almost exclusively provoked fun and play rather than aggression. Playing the dozens, Abrahams observed, provided a context for the boys to test one another in ways that explored sexual identity and thickened their young skin as a defense against the institutionalized hostilities they faced in the inner city. The dozens is the intellectual predecessor to rap and employs exaggeration and other signals of nonliteral meaning—rhyming, repetition.

Don't talk about my mother 'cause you'll make me mad.
Don't forget how many your mother had.
She didn't have one, she didn't have two,
She had eighty motherfuckers just like you.

I fucked your mother in a bowl of rice.
Two children jumped out shootin' dice.
One shot seven and one shot eleven.
God damn, them children ain't goin' to heaven.

Repetition is a classic element of the tease, and violates the rule of quantity. If a friend says you are a really good neck rubber, you blush with pride. If she says you are a really, really, really, really outrageously fantastic neck rubber, you are likely to bristle a bit, recall questionable massage techniques—the use of your elbows and your nose—you've experimented with, wonder what her point is, and rise to defend yourself.

Repetitive formulaic expressions rhythmically placed within social routines signal teasing. These linguistic acts are a reliable part of the quotidian life of healthy families. Parents have been known to short-circuit their children's mutinous reactions to their dinner with repetitive, formulaic expressions ("here's your dog food") to make light of, and preempt, their prickly objections.

We violate the rule of manner, or directness and clarity, in innumerable ways to tease. Idiomatic expressions—quirky nicknames and relationship-specific phrases—are a common element of teasing, zeroing in on idiosyncrasies and potentially problematic characteristics of the target. We violate the rules of manner with several vocal cues, including sing-song voice, loud, rapid delivery, dramatized sighs, and utterances that are either louder or quieter than preceding utterances. All of these acts deviate from the prosody of clarity and directness. And of course there is the wink, the very emblem of off-record indirectness. The wink violates the sincere and truthful orientation of direct, straightforward gaze, and recognizes an audience to the side, thus signaling that all is not what it appears to be.

With exaggeration, repetition, and idiomatic phrases, with elongated vowels and shifts in the speed and pitch of our delivery, with tongue protrusions, well-timed laughs, and expressive caricature of others, we violate the maxims of sincere communication, all in the service of teasing. We provoke, on the one hand, but artfully signal that nonliteral interpretations of the provocation are possible. We signal that we do not necessarily mean what we say, that our actions are to be taken in the spirit of play.

When we tease, linguist Herb Clark observes, we frame the interaction as one that occurs in a playful, nonserious realm of social exchange. When done with a light touch and style, teasing is a game, a dramatic performance, one filled with shared laughter that transforms conflicts—between rivals in a hierarchy, romantic partners, siblings finding separate spaces—into playful negotiations. It is in artful teasing that we lightheartedly provoke, to discern one another's commitments. It is with artful teasing that we convert many problems in social living to opportunities for higher *jen* ratios.

One of the clearest emblematic gestures—the tongue protrusion and finger wiggle near the head—signals that teasing is taking place. In this engraving of various Italian gestures, one sees the tease in the middle row—the thumb-to-nose finger wiggle.

POLITE ROARS AND CROAKS

The philosopher Bertrand Russell argued, "The fundamental concept in social science is power, in the same sense that Energy is the fundamental concept in physics." Power is a basic force in human relationships.

Power hierarchies have many benefits. Hierarchies help organize the collective actions necessary to gathering resources, raising offspring, defense, and mating. They provide heuristic, quick-decision rules about the allocation of resources and the division of labor (often favoring those in power). They provide protection for those involved (and peril to those outside the hierarchy).

Alongside their benefits, hierarchies are costly to negotiate. Conflicts over rank and status are very often a deadly affair. Male fig wasps have large mandibles that they put to deadly use in conflicts over mates and territory, most typically chopping each other in half. When several males find themselves on the same fig fruit, lethal combat quickly arises. One fig fruit contained 15 females, 12 uninjured males, and 42 damaged males who were dead or on their way to dying, with holes in the thorax and abdomen. Male narwhals use their drill-like tusks to conduct their negotiations over rank. In one hierarchy of male narwhals, over 60 percent of the males had broken tusks, and most had head scars or tips of tusks embedded in their jaws.

Given the enormous costs of negotiating rank, many species have shifted to ritualized battles. Displays of strength are exchanged in symbolic, dramatized form, and rank is negotiated through signaling rather than costly physical engagement. Red deer stags establish their rank in the autumn rut with roaring. The male who can roar louder and faster is assumed to be larger and stronger, and enjoys the ensuing evolutionary benefits (and, one hopes, the pleasures) of large harems of females. Harem holders will roar for hours on end, into the wee hours of the morning. They often lose weight in the process, to best their peers—all of which is a much better alternative than direct combat, injury, and an increased probability of death.

It is well known that many frogs and toads use the depth of their croaks to negotiate rank. Male frogs in one experiment were much less likely to attack another male when a deep croak was played by loudspeaker next to the pair. The deep vocalization, produced by large vocal cords, was assumed by both to be the signal of an exceptionally strong rival.

In humans, teasing can be thought of as the stag's roar or the frog's croak—a ritualized, symbolic means by which group members negotiate rank. Teasing is a dramatized performance clearly preferable to the obvious alternative—violent confrontations over rank and honor. Guided by this reasoning, former student Erin Heerey and I sought to capture teasing as a ritualized status contest. The issue we confronted was how to capture these brief status contests, so prevalent in the locker rooms and dugouts and keg parties of male youth, in the laboratory. Having people write narratives about their teasing experiences would miss the very heart of the tease—the nonverbal, off-record markers that give shape to the playfulness of the tease. We could have followed the formation of social hierarchies and the role teasing plays in naturalistic groups. Ritch Savin Williams had done this in a captivating study of boys' summer camps in the 1970s, and found, indeed, that ten- to twelve-year-old boys who were rising to the top of the hierarchy, like those dominant red deer, did indeed tease more to establish their elevated positions. But we wanted to capture the subtle, exceedingly brief nonverbal arabesques of teasing, those off-record markers, which required close-up videotaping.

In light of these interests, we developed the nickname teasing paradigm. Nicknames are a universal, linguistic marker of intimate relations that irrepressibly emerge in healthy marriages, friendships, joking relations between uncles and nieces, and work relations. Nicknames tend to home in on quirks, foibles, and deviant qualities of the target, but provoke in a loving way by violating the rules of literal communication (see examples in table below). Nicknames systematically involve exaggeration, repetition (alliteration), and metaphor (equating the individual with an animal or food, most typically). Nicknames are place holders for escapes to the world of play and pretense, where we can critique and mock in playful fashion without causing offense.

NICKNAMES FROM SPORTS AND POLITICS

MUHAMMAD ALI	THE LOUISVILLE LIP
JOE LOUIS	THE BROWN BOMBER
ROBERTO DURAN	NO MÁS
JAKE LAMOTTA	RAGING BULL
Y. A. TITTLE	THE BALD EAGLE
SHAQUILLE O'NEAL	BIG ARISTOTLE
KEVIN MCHALE	THE BLACK HOLE
JACK NICKLAUS	THE GOLDEN BEAR

LARRY JOHNSON	GRANDMA MA
BJORN BORG	ICE BORG
JOE BRYANT	JELLY BEAN
CHRIS EVERT	LITTLE MISS POKER FACE
KEN ROSEWALL	MUSCLES
JOHN ELWAY	MR. ED
JAROMIR JAGR	PUFF NUTS
KEITH WOOD	THE RAGING POTATO
WILLIAM PERRY	THE REFRIGERATOR
CHARLES BARKLEY	THE ROUND MOUND OF REBOUND
PAU GASOL	THE SPANISH FLY
ANTHONY WEBB	SPUD
GEORGE W. BUSH	BUSH 43, DUBYA, THE SHRUB, UNCURIOUS GEORGE
BILL CLINTON	THE COMEBACK KID, THE FIRST BLACK PRESIDENT, SLICK WILLIE
RICHARD NIXON	TRICKY DICK, IRON BUTT, THE MAD MONK
GEORGE WASHINGTON	THE OLD FOX, THE FARMER PRESIDENT
JOHN ADAMS	BONNY JOHNNY, YOUR SUPERFLUOUS EXCELLENCY, HIS ROTUNDITY
ABRAHAM LINCOLN	HONEST ABE, THE ILLINOIS APE, THE GREAT EMANCIPATOR

Our nickname paradigm was to present participants with two randomly generated initials—A. D. or T. J. or H. F. or L. I. Participants then generated a nickname for the eventual target of the tease based on those letters, as well as an accompanying story—fact or fiction—that justified the nickname. We encouraged our participants not to worry about profanities or lewdness, notwithstanding their being videotaped; we said we weren't going to post their videotapes on the Internet or send them to their grandmothers.

To examine how teasing functions as a status contest, I enlisted an honors student, Mike Bradley, a bright young member of a fraternity at the University of Wisconsin-Madison. The fraternity housed seventy-five members, tightly packed into a grand old mansion on Lake Mendota. With Mike's help, we brought groups of four fraternity members—two high-status "actives," who were established members of the group, and two new, low-status pledges—to the laboratory to tease one another with our newly minted nickname paradigm. They came in October, just as the group was forming its new identity amid the falling leaves and darkening trees in the upper Great Lakes fall. Fraternity members are notorious for their teasing. When told that they were participating in a study of teasing, the high-status actives licked their chops, and the low-status pledges dropped their gaze and shook their heads with a knowing smile, sensing what was coming.

The teasing flowed out of the mouths of the fraternity brothers in bursts of profane, cartoonish poetry, resembling the ritualized insults observed across history and culture. The great satirist Rabelais described nicknames used in a quarrel between cake bakers and shepherds, who playfully violated Grice's maxims through use of exaggeration ("shit-a-beds"), repetitive alliteration ("crazy carrot-heads," "mincing milksops"), and oblique metaphor ("poor fish"):

> babblers, snaggle-teeth, crazy carrot heads, scabs, shit-a-beds, boors, sly cheats, lazy louts, fancy fellows, drunkards, braggarts, good-for-nothings, dunderheads, nut-shellers, beggars, sneak-thieves, mincing milksops, apers of their betters, idlers, half-wits, gapers, hovel-dwellers, poor fish, cacklers, conceited monkeys, teeth-clatterers, dung-drovers, shitten shepherds.

Our participants resorted to their own earthy patois, generating nicknames like "turkey jerk," "little impotent," "human fly," "anal duck," "heffer fetcher," and "another drunk." Systematically contained in the teasing were admonitions about transgressions that could unsettle the fraternity. There were numerous references to excessive drinking, and roughly a quarter of the teases made reference, often a bit incongruous in the context of the story being told, to the target's genitalia. One story directed at a low-status pledge called him "Taco John" and revealed to the audience how this pledge had gotten so wasted on eighteen shots of Bacardi that at a late-night feast at the fast-food spot, Taco John's, he disappeared, and was found passed out in the bathroom, near a toilet, holding his genitals. The fraternity members were notifying each other about moral boundaries: Don't get too drunk, and keep your genitals to yourself.

More in-depth coding focused on the provocative aggression of the tease—easy to detect in terms of its aggressive and humiliating content—as well as the off-record markers that render the tease less biting. These include shifts in voice, funny facial expressions, laughter, and use of metaphor and exaggeration. After the months it took to code these thirty- to forty-second bursts of teasing, we found a clear story about status and teasing, represented in the figure below. High-status actives teased everyone, in particular low-status pledges, in aggressive, provocative fashion, putting them in their place. Low-status pledges flattered their new high-status brothers, recognizing the elevated rank of the actives. With sharp teeth, however, they went after the other low-status pledges, no doubt jousting for an edge. We also found that more popular pledges, some of whom had great charisma and charm, and were clearly and quickly on the rise, were teased in more flattering fashion. Within a couple of weeks of the group's formation at the start of the academic year, thirty-second teases were clearly demarcating rank.

If one were to study transcripts of the spoken words of these teasing battles, one would have expected affront, aggression, and perhaps a thrown punch or two. Instead, these groups of four fraternity members laughed in hysterical unison. They patted each other on the back and pushed each other playfully. They growled and pointed in mock aggression. In the briefest of instances they looked

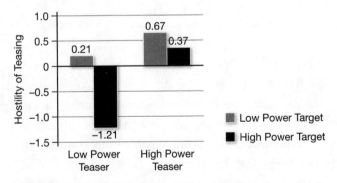

High-power fraternity members tease in more hostile fashion than low-power teasers, who show reserve when teasing high-power members.

into each other's eyes. In fact, in my twenty years of science, which has involved thousands of participants, this study produced two anomalies. I had members of the fraternity who had not been in the study call me on my office phone to ask whether they too could be in the study. And I had a couple of participants ask if they could be in the study a second time (which prompted a rather dull lecture on my part about how science requires independent observations of single participants).

Notwithstanding the degrading nicknames and humiliating tales of perverse sexuality and exposed genitals, fraternity members indicated that they thought more highly of the three guys they had just teased and been teased by than the other members of the fraternity. My coding of the laughter and embarrassment revealed how. The more a teaser and target fell into antiphonal or shared laughter, the more they liked each other. The more the target blushed and showed subtle signs of embarrassment (the gaze aversion, the face touch, the head movement down and away), which often ended in conciliatory eye contact between teaser and target, the more the teaser liked the target. Teasing, when done well, provides a platform for negotiating conflict-laden relationships—in this case, positions of rank in a hierarchy—in playful, friendly fashion. And on that platform of playful provocation, teasing evokes brief bursts of emotion—shared laughter, the urge to reconcile at another's embarrassment—that move individuals toward greater *jen*.

MERRY WAR

A few years ago I was vacationing with my family on a cool, white-sand beach near Monterey, California. While we were building sand piles, bodysurfing small waves, and looking for sand crabs in the shallows of the frothy surf, a group of Mexican-American teenagers descended upon our peaceful place in the sun. Clad in the blue trousers and pressed white shirts of their Catholic school, they approached the beach under the watch of their teacher in gender-segregated, orderly, single-file lines. Once situated, with the sound of the surf and away from their teacher, who was enjoying a well-deserved moment of repose, they broke into a bedlam of teenage teasing.

The five boys and six girls were like molecules bound together by the attractive forces of teasing. There was a continuous stream of pinching, head rubbing, poking, squeezing, name calling, howling, and laughing. As rhythmic as the sound of the ocean, two boys would grab a girl, hold her by her arms and legs, and dangle her over the ebb and flow of the waves. Three demure girls sneaked up on a boy and tried to tug his low-hanging pants down. He forcefully countered with fistfuls of sand. Water was dripped on necks. Sand was pressed into others' pants. Seaweed dangled in front of faces. Dog piles occasionally broke out. In a surprise attack one girl managed to nearly drop a dead crab down a boy's pants. When their teacher called them back to the bus, they regained their composure and left in two lines, one of boys, the other of girls.

As they departed, my daughter Serafina, then 5, asked me: "Why did that girl put the crab in the boy's pants?"

"Because she likes him," I responded.

This left Serafina dumbfounded. So I mumbled something unintelligible about how we actually tease people we like, Grice's maxims, and the playful realm of off-record language, and told her that we often mean the opposite of what we say or do. What Serafina took from viewing this teenage drama, I hope, is wisdom about the invaluable place teasing has in intimate relations.

There is no relation more vital to the survival of our species than the intimate bond. There is no relationship more fraught with conflict or more fragile. Our ultravulnerable, big-brained offspring require more than one caregiver to survive, thus binding us into long-term caretaking relations, which have no parallel in our close primate relatives. And from their moment of inception to their end, intimate relations roil with conflict, sacrifice, and matters to negotiate. In early stages of intimacy, sexual strategies—interests in short-term exchange or long-term devotion—require navigation. As children, housework, and mortgages arrive, partners can feel like beleaguered managers of a halfway house, moving from one crisis to the next. Intimate life is a "merry war," as noted by Leonato in *Much Ado About Nothing*.

So we turn to teasing to solve many problems of intimate life. We tease to flirt, to discover others' affections and sexual interests. Monica Moore surreptitiously observed teenagers at a mall, and found their moblike meanderings to be punctuated by bursts of teasing. Young boys and girls would routinely veer into each other's orbits to pinch, tickle, poke, and squeeze, creating, of course, opportunities for physical contact and brief mutual gaze—so highly regulated during the self-conscious teenage years. For young teens, teasing is a drama in which telltale signs of attraction—the blush, the lip pucker, the mellifluous, "voiced" laugh, the gaze that lasts beyond the .45-second eye contact that defines more formal social exchange—are sought with hormonally charge voracity amid the razor-sharp surveillance of peers. Teasing is an entrance into a playful world in which potential suitors can test and provoke one another. Were contemporary teens more restricted in their physical contact, they would resort to the war of words that marks the first encounter of Benedick and Beatrice in *Much Ado About Nothing*, sure signs that they are to fall in love:

BEATRICE   I wonder that you will still be talking, Signior Benedick. Nobody marks you.

BENEDICK   What, my dear Lady Disdain! Are you yet living?

BEATRICE   Is it possible disdain should die while she hath such meet food to feed it as Signior Benedick? Courtesy itself must convert to disdain, if you come in her presence.

BENEDICK   Then is courtesy a turncoat. But it is certain I am lov'd of all ladies, only you excepted; and I would I could find in my heart that I had not a hard heart, for truly I love none.

BEATRICE   A dear happiness to women! They would else have been troubled with a pernicious suitor. I thank God and my cold blood, I am of your humor for that. I had rather hear my dog bark at a crow than a man swear he loves me.

BENEDICK   God keep your ladyship still in that mind! So some gentleman or other shall scape a predestinate scratch'd face.

BEATRICE	Scratching could not make it worse, an 'twere such a face as yours were.
BENEDICK	Well, you are a rare parrot-teacher.
BEATRICE	A bird of my tongue is better than a beast of yours.
BENEDICK	I would my horse had the speed of your tongue, and so good a continuer. But keep your way, a God's name; I have done.
BEATRICE	You always end with a jade's trick, I know you of old.

With age, as sexual partners move in with one another, sharing the conduct of everyday life, partners develop their own personal idiom, including provocative nicknames and teasing insults. Partners develop teasing insults about each other's sexual proclivities, their bodily functions, their sleep habits, their manner of eating, the anachronistic hairstyle they dogmatically prefer—all threats to connubial bliss. These teasing insults mark partners' quirks and foibles as deviant, and problematic if carried to extremes, but endearing foibles nonetheless, uniquely appreciated by the partner. Studies of married partners find that partners with a richer vocabulary of teasing insults are happier, and enjoy a better long-term prognosis.

The teasing of romantic partners—nicknames, ritualized insults—not only signals their unique intimacy together. It also provides a realm of pretense in which the two can playfully negotiate their conflicts. To explore this possibility, I had couples who had been together for several years tease each other with the nickname task. Roughly a quarter of the nicknames they generated involved universal metaphors of love—references to their partner as pieces of food (apple dumpling) or small animals. Once again, a nod to the benefits of teasing in the merry war of intimacy: The more satisfied couples were more adroit at teasing; they were more likely to use off-record markers in their fifteen seconds of teasing—exaggerations, repetitions, mimicry, playful intonations, shifts in pitch, tongue protrusions, contorted facial expressions. They had developed a nonliteral, off-record dimension to their intimate life, one which they could readily transport to, to enjoy the levity, antiphonal laughter, and lightness of the realm of pretense. The playfulness of their fifteen-second teasing, we additionally found, predicted how happy the couples were six months later.

In another study we examined the precise instances in which partners criticized each other. We identified specific moments as they haggled over a serious conflict in their relationships—money issues, future commitment, infidelities, questions about how they spend time together. Much of the time partners delivered the criticism in well-reasoned, on-record, literal prose that would satisfy any juror or local rhetorician. Other times partners teased to criticize—with exaggerated claims, a ritualized insult, a playful imitation of the other, a nickname, mock anger or frustration. They delivered the same content—some provocative criticism, for example, about the partner's bevy of ne'er-do-well friends or tendency to spend too much money—but did so in nonliteral fashion, employing off-record markers to indicate a nonserious side to what they were saying. Couples who playfully teased, as opposed to resorting to direct, cogent, but ultimately hackle-raising criticism, felt more connected after the conflict, and trusted their partners more. The playful dramatizing of conflict is an antidote to toxic criticism that can dissolve an intimate bond. Teasing is a battle plan for the merry war.

THE PLAYGROUND PARADOX

On April 20—the birthday of Adolf Hitler—1999, Eric Harris and Dylan Klebold entered Columbine High School in Littleton, Colorado, armed with semiautomatic weapons and a hit list, and proceeded to kill twelve of their fellow students, a teacher, and then themselves. In a burst of soul searching, Americans raised muted questions about guns, games, and drugs. Harris and Klebold had made over ninety-nine bombing devices, and had an astonishing collection of firearms, hidden from adults. They were avid players of Doom, one of many violent videogames known now, thanks to research by Craig Anderson and Brad Bushman, to short-circuit compassionate tendencies and to amplify aggression. After going on Zoloft and then Luvox for bouts of depression, Harris became increasingly prone to homicidal and suicidal thoughts.

In the immediate aftermath of the Columbine shootings, I received a call from a counselor who had been working with children there. Harris and Klebold had been bullied by the jocks at

Columbine High. This bullying, some suggested, had been condoned by administrators at the school—a fact that led to many zero-tolerance policies toward bullying in schools. Her deep concern, in reading my research, was that I was condoning bullying by saying that teasing is good.

The simple answer to her query is that the heart of bullying has nothing to do with teasing. What bullies largely do is act violently—they torment, hit, pin down, steal, and vandalize. This has little to do with teasing.

The more subtle matter we confronted is the paradox of the playground. Scan a playground of any grammar school for fifteen minutes and you'll see the full spectrum of teasing, its lighter, playful side as well as its darker versions. Children have an instinct for teasing. It emerges early (one British psychologist observed a cheeky nine-month-old mocking her grandmother's snoring with a delightful imitation). As with adults, teasing can instigate and mark deep friendship. At the same time, teasing can go horribly awry. The teasing of children with obesity problems, for example, has been found to have lasting pernicious effects upon the target's self-esteem.

What separates the productive tease from the damaging one? Data from our studies yielded four lessons about when teasing goes awry, lessons that can be put to use on the playground or in the office. A first is the nature of the provocation in the tease. Harmful teasing is physically painful and zeroes in on vulnerables aspects of the individual's identity (for example, a young man's romantic failures). Playful teasing is less hurtful physically, and thoughtfully targets less critical facets of the target's identity (for example, a young man's quirky manner of laughing). The literature on bullies bears this out: Their pokes in the ribs, noogies, and skin twisters hurt, and they tease others about taboo subjects. Not so for the artful teaser, whose teasing is lighter and less hurtful, and can even find ways to flatter in the provocation.

A second lesson pertains to the presence of the off-record markers—the exaggeration, repetition, shifts in vocalization patterns, funny facial displays. In studies of teasing we have found that the same provocation delivered with the wonderful arabesques of our nonliteral language, the off-record markers, produced little anger, and elevated love, amusement, and mirth. The same provocation delivered without these markers mainly produced anger and affront. To sort out the effective tease from the hostile attack, look and listen for off-record markers, those tickets to the realm of pretense and play.

A third lesson is one of social context. The same action—a personal joke, a critical comment, an unusually long gaze, a touch to the space between the shoulder and neck—can take on radically different meanings depending on the context. These behaviors have different meanings when coming from foe or friend, whether they occur in a formal or informal setting, alone in a room or surrounded by friends. Critical to the meaning of the tease is power. Power asymmetries—and in particular, when targets are unable through coercion or context to respond in kind—produce pernicious teasing. When I coded the facial displays of the twenty-second bursts of teasing in the fraternity study, amid the laughter and hilarity I found that over 50 percent of low-power members showed fleeting facial signs of fear, consistent with the tendency for low power to trigger a threat system—anxiety, amygdala hyperreactivity, the stress hormone cortisol—which can lead to health problems, disease, and shortened lives when chronically activated. Bullies are known for teasing in domineering ways that prevent the target from reciprocating. Teasing in romantic bonds defined by power asymmetries takes the shape of bullying. The art of the tease is to enable reciprocity and back-and-forth exchange. An effective teaser invites being teased.

Finally, we must remember that teasing, like so many things, gets better with age. Starting at around age ten or eleven, children become much more sophisticated in their abilities to endorse contradictory propositions about objects in the world—they move from Manichaean, either/or, black-or-white reasoning to a more ironic, complex understanding of the world. As a result, as any chagrined parent will tell you, they add irony and sarcasm to their social repertoire. One sees, at this age, a precipitous twofold drop in the reported incidences of bullying. And this shift in the ability to understand and communicate irony and sarcasm should shift the tenor of teasing in reliable fashion.

To document this, we created an opportunity for boys at two different developmental stages to taunt one another at a basketball camp. The camp was run by my former student, aptly named John Tauer—a former Division III star small forward at the University of St. Thomas. During the camp's morning drills, two boys, who did not know each other but were matched according to their basketball skills, were called over to play the "pressure cooker." In this drill each camper was to try to make a free throw with the game on the line: If he made the shot, his team would win, and if he missed, his team would lose. Each camper's partner was to act like a fan for fifteen seconds prior to the critical shot, much as NBA players must perform amid the taunting coming from the stands. In the taunt condition, the fan was given the task of messing with the shooter's mind for fifteen seconds, to try to make him miss. In the cheer condition, the fan was to say encouraging things, to bolster the spirits of the shooter. The fan acted this out in a taped square that measured two feet by two feet and was located at the end of the left side of the free-throw line, a couple feet away from the shooter.

Sure enough, our subjects acted like besotted NBA fans. The taunters made more than ten times as many harsh gestures as those cheering their peers on. The taunters pointed and stuck out their tongues, they growled and snarled, they glared like alpha apes. The cheering fans, in contrast, took their cue straight out of a self-esteem handbook: They were five times as likely to shout encouraging things, to clap, and to cheer the shooter on. And there, in the fifteen-second episode of taunting, and only in the taunting, we saw the exquisite use of off-record markers. The taunters, and only the taunters, shifted their pitch, taunting in very high or low tones, they resorted to repetition ("you're gonna miss, you're gonna miss, you're gonna miss), and they used basketball metaphors ("brick" "choke") that were absent in the more sincere cheer condition.

The fourteen- to fifteen-year-old boys taunted with just as much hostile behavior—the finger points and harsh vocalizations—as the ten- to eleven-year-old boys, but the older boys' taunting was accompanied much more frequently by off-record markers—repetition, shifts in vocalizations, metaphors—that signaled the presence of play. And to good effect. Our older campers, deft taunters that they were, were more likely than the older boys in the cheer condition to report their partner as a new friend. In the aftermath of the taunt and the shot, whether a make or a miss, the two boys' laughter would become intertwined. They would nudge each other in the shoulder or get into friendly headlocks. They would push each other gently and sometimes walk to the next station with an arm draped on the new friend's shoulder.

PARALLEL PLAY

There are still many mysteries to Asperger's Syndrome, a condition defined by typically developing or accelerated language and cognitive abilities but profound difficulties in the social realm, and in connecting with and understanding others. Why is it three to four times more common in young boys than girls? Is it a disorder? Or should it be thought of as just another hue in the human spectrum?

There is no mystery to the heart of the condition, as revealed in the brilliant essay by music critic Tim Page, who has lived his life with Asperger's. It is a condition defined by early single-minded preoccupations—in Page's case, maps of towns in Massachusetts, obituaries, memorizing most of the 1961 edition of the *World Book Encyclopedia*, and the music of Scottish comedian Harry Lauder (he publicly declared his contempt for the Beatles in his school newspaper). It is manifest in an unusual social style—often defined by a monotone voice, a lack of eye contact, a revulsion to touch, and fearless social oddities (Page liked to wear rabbit's feet in each buttonhole of his shirt). It is no wonder that Page found *Emily Post's Etiquette* an epiphany, a step-by-step manual for entering into the social complexities of the human race.

At the same time, the disinterested disregard for others can yield prodigious talents. Hans Asperger, the Viennese pediatrician who helped chart the nature of the condition, observed, "For success in science or art, a dash of autism is essential." For Page, it led to deep insights into music. After hearing Steve Reich's *Music for 18 Musicians* in 1976, he was catapulted into a five-year study of minimalism in contemporary music that led to his career as a music critic. He observed that Reich had achieved the musical equivalent of imposing a frame upon a moving river.

At childhood birthday parties Page felt deeper empathy for the piñata than his sugar-juiced peers. In his teens, when in the close confines of interested young girls, he would chatter on without making eye contact. Later he confessed to making love like the Tin Man. About his understanding of others, Page writes: "I am left with the melancholy sensation that my life has been spent in a perpetual state of parallel play, alongside, but distinctly apart from, the rest of humanity."

What proves to be difficult for Asperger's children are the tools of social connection, all those contributors to the numerator of the *jen* ratio—eye contact, gentle touch, the understanding of others' minds, embarrassment or love, imaginative play with others, greeting smiles with smiles, antiphonal laughter. And teasing, as revealed in a study I conducted with my friend and colleague Lisa Capps. If teasing is a dramatic performance, one that requires nonliteral language, where affections, conflicts, commitments, and identities are playfully negotiated, this should be particularly difficult for Asperger's children. They have difficulties in imaginative play, pretense, taking others' perspectives, and the elements of the tease, in particular nonliteral communication.

In our study we visited the homes of Asperger's children and their mothers, as well as the homes of comparison children and their mothers. We asked the children to recount experiences of teasing. We then had them tease each other with the nickname paradigm. Our children were 10.8 years old, on average—the very age that children's capacities for multiple representations and irony come on line and teasing transforms into a pleasurable social drama. Our comparison children described experiences of teasing that had many positive flavors, in which they navigated the connections and moral notions of preteen life. The Asperger's children, in contrast, recounted experiences that were largely negative, and made little reference to connection and community. When we coded the brief teasing exchanges between parents and child, we found out why. Asperger's children were just as hostile in their teasing of their mothers as comparison children, but they showed none of the nonliteral gems of an artful tease—exaggeration, repetition, prosodic shifts, funny facial expressions, imitations, iconic gestures, metaphor. These difficulties with the tease, we also found, could be attributed to the child's difficulties with taking others' perspectives.

In summing up his life with Asperger's, Page reflects:

> I cannot pretend that Asperger's has not made much of my existence miserable and isolated (how *will* I get to sleep tonight?). I hope that young Aspies, informed by recent literature on the subject, will find the world somewhat less challenging than I have.

As one of our young Asperger's children said: "There are some things I don't know so much about. . . . Teasing is one of them." Absent teasing, the Asperger's child misses out on a layer of social life, of dramatic performances where affections are realized, roles are defined, conflicts are hashed out, all in the lighthearted rhetoric of nonliteral language. They miss out on what teasing gives us: shared laughter, playful touch, ritualized reconciliation, the perspective of others—a life beyond parallel play.

9

Touch

FOR THE PAST FIFTEEN YEARS His Holiness the Dalai Lama (HHDL) has been engaging scientists and Tibetan Buddhists in dialogues about the interface between the science of the human nervous system and the 2500-year-old tenets of Tibetan Buddhism. The intention is to uncover commonalities and differences in how these two ways of knowing arrive at claims about the nature of mind, emotion, and happiness.

I have participated in two Buddhism-Science panels, one at MIT and the other through the Dalai Lama Peace Center in Vancouver Canada. Each event evoked out-of-body feelings of the unreality of a wedding day. There was the swirl of the 200 photographers who track HHDL's every bow, smile, laugh, attentive head nod, finger pointing rhetorical flourish, cough, and sneeze. At MIT, bomb-sniffing dogs circled the space-age auditorium, sniffing under chairs and behind posters protesting for a free Tibetan state. Stone-faced secret service agents, mumbling into collar microphones, stood in dark suits positioned in perfectly proportioned geometric arrangements to protect the space that HHDL moved through. Outside the MIT auditorium at 6 AM hundreds of meditators, peacefully arrayed in rows of upright torsos and braided hair, sat in contemplative anticipation, occasionally awakened by the strike of gongs. The dozens of volunteers who made these events possible, long time

His Holiness the Dalai Lama greets the author.

practitioners of Tibetan Buddhism from the west, uniformly spoke of being touched by HHDL. Literally touched. One recalled clasping his hands years ago in New York. Another recalled a brush during a bow in Daram Sala India, where HHDL resides. Still another recalled his hand on a shoulder at a reception following a talk. They could remember the precise instant of the touch, the warm feeling that rippled through their bodies, and the lasting change this contact introduced into their lives. Often the act of recalling the touch produced bright eyes, a flush, a head tilt, tearing, and an intimate but remote look in the eyes.

On the stage in Vancouver before our dialogue, HHDL entered stage left and proceeded to greet the four panelists with his customary bow and clasped hands. The sighs, tears, appreciative head nods, goose bumps, and embraces of the 2,500 people in the audience produced a crackling ether that filled the art deco auditorium. I was the last panelist for HHDL to approach. From eighteen inches away I came into contact with HHDL. Partially stooped in a bow, he made eye contact with me and clasped my hands. His eyebrows were raised. His eyes gleamed. His modest smile was poised near a laugh.

Emerging out of the bow and clasped hands, he embraced my shoulders and shook them slightly with warm hands.

As he turned to the audience, I had a Darwinian spiritual experience. Goose bumps spread across my back like wind on water, starting at the base of my spine and rolling up to my scalp. A flush of humility moved up my face from my cheeks to my forehead and dissipated near the crown of my head. Tears welled up, along with a smile. I recalled a saying of HHDL's:

> At the most fundamental level our nature is compassionate, and that cooperation, not conflict, lies at the heart of the basic principles that govern our human existence.

For several weeks after I lived in a new realm. My suitcase was missing at the carousel following the plane flight home—not a problem, I didn't need those clothes anyway. Squabbles between my two daughters about the ownership of a Polly Pocket or about whose back-bending walkover best matched the platonic ideal—no bristling reaction on my part, just an inclination to step into the fray and to lay out a softer discourse and sense of common ground. The frustrated person behind me in the line in the bank, groaning in exasperation—no reciprocal frustration, no self-righteous sense of how to comport oneself in more dignified fashion in public; instead, an appreciation of what deeper causes might have produced such apparent malaise. The people I saw, the undergrads in my classroom, parents at my daughters' school, preschool teachers walking little groups of three-year-olds in hand-holding chains around the streets of Berkeley, those parallel parking their cars, recyclers picking up cans and bottles, the homeless shaking their heads and cursing the skies, people in business suits reading the morning paper waiting for a carpool ride, all seemed guided by remarkably good intentions. My *jen* ratio was approaching infinity.

I wish I could do full-body fMRIs of people's nervous systems in settings through which HHDL moves. If I could, I would find that his touch produces the colorful activation of goodwill in the brain and body. HHDL's vocabulary of touch is as precise and imaginative as a chess master's representation of the possibilities on the 64 squares of the chessboard. As we concluded our panel in Vancouver, HHDL tickled Paul Ekman in the ribs. In the midst of a discussion about neural plasticity, HHDL squeezed neuroscientist Richie Davidson's earlobe. In his deep, bowed greetings with other Tibetan monks, HHDL rubs the corner of his head against that of the other monk, triggering laughter. HHDL is known to fall to the floor and wrestle with Desmond Tutu, as though the two were preteen brothers. HHDL's genius at touch is a window into an ancient communicative system by which we can alter others' *jen* ratios, spreading health and happiness to others.

VIRAL *JEN*

In John Huston's *The Treasure of the Sierra Madre,* Humphrey Bogart, tired of bumming smokes and meals off expatriate Americans in Tampico, Mexico, encamps with two other down-on-their-luck prospectors in the arid mountains of the Sierra Madre. They are in search of gold. As their bags of gold dust mount in weight and number, the three men confront an ancient evolutionary problem—how to build and maintain trust between nonkin. In their high chaparral camp, the opportunities for exploitation are infinite—a quick escape with the gold during the heavy sleep following a day of moving dirt, a silent murder in a desert ravine, an alliance of two ganging up on the third. In the face of the pull of exploitative self-interest, the band of desperate prospectors hold tight. They are bound together in cooperative spirit by enthusiasm, camaraderie, the reverie of the meals and clothes and farms and white picket fences they envision enjoying with their newfound fortunes, and the laughter, banter, backslapping, and firm handshakes of men cooperating.

When a mine shaft collapses on Bogart, he suffers a blow to his head. Like Muybridge, his mind shifts to the orientation of dog-eat-dog survival, and he leads the group into a nightmarish battle of distrust and exploitation. Ambiguous actions out in the desert—a partner claims to be searching under a rock for a Gila monster—appear to cynical eyes to be attempts at searching for the other's

hidden bags of gold. The language of friendship—"buddy,"
"friend," nicknames—shifts to the sharp, impersonal tones of last
names. Suspicions about the imagined interests of others escalate
into gun-pointing confrontations.

This high mountain drama parallels a central dynamic in the evolution of cooperation. How might cooperation, kindness and *jen*
emerge in social groups composed of individuals who are better off
pursuing self-interest? As we learned from the lessons of the tit-for-
tat, an answer is found in the contagious goodness hypothesis. Simply put, cooperation and kindness will take hold in groups when *jen*
becomes viral, when individuals can readily signal their kind intentions to others and evoke similar inclinations. In this fashion, people are less likely to experience the costs of being generous to
competitive individuals, and more likely to enjoy the fruits of
mutual cooperation—reciprocal trades of resources, sharing in
parental care, and so on.

Jen becomes viral through behaviors that spread goodness from
one individual to the next, thus setting in motion reinforcing, reciprocal cooperation. These behaviors would need to be powerful and
fast, to counteract the mind's automatic tendency to perceive
threat, danger, and competition in nearby, fast-moving bundles of
self-interest—namely other humans. These behaviors would need
to operate on the bodies of others, to shift the nervous system away
from its potent, trigger-happy fight/flight tendencies toward a physiological profile more conducive to cooperation and kindness.
These contagious behaviors would need to be easy to use and readily adaptable to the close proximity of the daily interactions of our
hominid predecessors. These contagious behaviors, as signals of
cooperation and trust, would need to be easy to perceive and hard
to feign.

The scientific study of touch reveals the tactile modality to be
an ideal medium in which individuals spread goodness to others.
We can readily put touch to use in the close encounters of group
living—when negotiating close spaces, working together, flirting
amid rivals, playing around, or allocating scarce resources. Touch
triggers biochemical reactions in the recipient—activation of the
orbitofrontal cortex and deactivation of the amygdala, reduced

stress-related cardiovascular response, and increased neurochemicals like oxytocin—all of which promote trust and goodwill between individuals. Touch, my studies show, is the primary language of compassion, love, and gratitude—emotions at the heart of trust and cooperation. To understand why touch can make *jen* viral, we must first look to evolution of the largest organ of the body—the epidermis—and that five-digit wonder, our hand.

SKIN AND HAND

In humans, shifts in the morphology of our organs of communication have emerged with developments in our ultrasociality. So it is with touch: Evolutionary shifts in our skin and hands have led to a central role of touch in our ultrasocial relationships. A first big shift was the loss of most of our body hair—we became, in Desmond Morris's famous phrasing, naked apes. Why? You may be tempted by the aquatic ape hypothesis—that for a period of our evolution we actually lived largely in the water, thus losing our hair, as other water-bound mammals such as dolphins and whales did. As readily as this hypothesis appeals to our love of lolling about in the water on hot summer days and our sense of communion with whales and dolphins, it makes little evolutionary sense. As Nina Jablonski has argued in her book *Skin*, water holes on the African savannah were highly dangerous places, brimming with quick-striking predators, all the more so to a species not terribly adroit in the water, as in the case of our hominid predecessors. Had we been aquatic apes, we wouldn't have done very well in the game of survival.

The actual explanation for our hair loss is less flamboyant but eminently more sensible—we lost our hair in hominid evolution for purposes of thermoregulation. A thick carpet of body hair would have been dangerously hot on the savannah—the locale of our early evolution. In this hot, arid environment, we increasingly relied on the rich network of sweat glands distributed throughout the skin to keep ourselves cool. These glands function more effectively in the absence of hair.

One by-product of this shift toward hairlessness is that our skin evolved into a most remarkable interface between our inner and outer worlds. Human skin is the largest of our organs, weighing six pounds and covering eighteen square feet. Its distinct layers house a veritable industrial zone of biological factories accomplishing several functions essential to human survival. A rich network of blood vessels, sweat glands, and hair follicles and surrounding muscles lie under the skin. There are cells producing proteins called keratins, which account for the strength and resilience of the skin. Cells known as immigrant cells move into the skin during development from other parts of the body and accomplish three tasks. Melanocytes produce the skin's pigment, melanin, which protects our bodies from the dangers of ultraviolet rays. Langerhans cells are part of our immune system, and represent our body's first response to viruses and bacteria. Finally, Merkel cells reside at the ends of sensory nerves in the skin and respond to touch. Some of these cells, in particular in the arm, face, and leg, appear to respond to slow, light touch, and may be involved in the release of opioids trigged by contact from others. The skin is our protection against harmful physical agents—sharp branches, ultraviolet rays, bacteria and viruses—in the external world. As important as the skin is to keeping the bad stuff out, it is vital to bringing the good stuff in.

Just as critically, evolutionary changes in the morphology of the human hand likewise facilitated the development of the tactile language of emotion. As humans began to walk upright, the hand

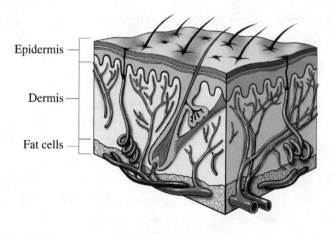

The layers of the skin

changed dramatically. We acquired the opposable thumb—the morphological darling of many evolutionists. We also developed more dexterous fingers. Chimp thumbs are much shorter, in relation to the rest of their hand, than the human thumb. Humans, unlike chimps or bonobos, became able to make precision grips between thumb and forefinger, and power grips using the entire hand. These shifts in the morphology of the hand, most obviously, allowed our hominid predecessors to emerge as the first complex toolmakers in primate evolution, fashioning sophisticated arrowheads, clothes, baskets, and so forth. In the process, we developed profoundly expressive hands. Our hands allowed us to point with precision, a critical part of the child's emergent understanding of the referential quality of language: Words refer to things. With the refined acrobatics of our hands and fingers, we learned to signal different objects and states with what are known as emblems—gestures that translate to words. With our hands we learned to convey internal states with specific patterns of touch.

The skin and hand evolved to enable adaptive response to heat and for tool use. Alongside these pragmatic gains, our hominid predecessors evolved a communicative system that became central to how humans form and maintain bonds. Most obviously, the skin is the platform for intimacy and sexual relations. It is a medium in which individuals in conflict channel aggression, through pinches, pokes, prods, and punches. We soothe and reassure with hands on skin. The skin and touch are a central medium in which the goodness of one individual can spread to another, resulting in high *jen* ratios as the primary orientation in groups.

CONTACT HIGH

Faith healers have been in human society since at least the time of the classical Greeks and Romans. Central to their repertoire of healing skills is touch. Recent neuroscience suggests that, at least in the use of touch, faith healers may have been on to something. Scientists have learned that touch is a basic reward, as potent as the sweetest of summer peaches or the scent of blooming jasmine. The progenitor of this view is Edmund Rolls, of Cambridge University, who has studied the orbitofrontal cortex (OFC), which, as we will recall, was damaged in Eadward Muybridge during his fateful stagecoach accident in east Texas. Rolls's thesis is that this region of the brain processes information about basic rewards, which help individuals navigate their physical and social environments, acting in ways, presumably, that bring about more rewarding social encounters, more nutrition-rich searches for foods, and so on. He has found that sweet tastes and pleasing smells stimulate activity in the OFC, in particular in hungry animals. But there is more. He has also documented that the simple touch of the arm with a soft velvety cloth activates the OFC, so important to our understanding of how to obtain rewards. This is a remarkable finding: Touch (the right kind, of course) is as powerful and immediate a reward as chocolate or the scent of Mother to an infant. Touch is a primary color in the color scheme of pleasure, wired deep into our nervous systems.

Further scientific studies found that touch—again, the right kind—sets in motion a cascade of rewarding biochemical reactions. For example, in one study participants received a fifteen-minute Swedish massage, the type that is stock in trade at spas and now a much-appreciated service at more forward-looking airports. While participants' Merkel cells were being pleasingly pressed, their blood was drawn. A quick neck rub to the shoulders by a stranger triggered the release of oxytocin, a neuropeptide that promotes oceanic feelings of devotion and trust. Other studies have found that massage (like Prozac) increases levels of the neurotransmitter serotonin, and that it reduces levels of the stress hormone cortisol. Touch also appears to release endorphins in the recipient—a natural source of pleasure and pain relief.

Touch, then, triggers activation of the orbitofrontal cortex and the release of oxytocin and endorphins—biological platforms of social connection. Just as importantly, recent studies of maternal behavior in rats suggest that the act of touching is physiologically rewarding for the toucher. Rat mothers devote a great deal of time to licking their rat pups and coming into nose-to-nose physical contact with their offspring. Recent studies have found that rat mothers who lick their pups a lot, who touch their offspring a great deal, get surges of dopamine upon physical contact. Dopamine is a neurotransmitter that is involved in the pursuit of rewards; it underlies our experience of sensory pleasure. When we touch, the implication is, we get a burst of pleasure as well.

Thankfully, the benefits of touching are not limited to rat pups. Scientists have found, for example, that depressed mothers who are encouraged to touch their infants regularly and to massage them experience reduced symptoms of depression and begin to play more with their children. Elderly individuals who volunteer to give massages to infants report reductions in anxiety and depression and enhanced well-being.

Humans, then, are blessed with an inexhaustible resource of rewards—touch. Through millisecond touches in our daily living, we can provide pleasure, reward, and encouragement to others. Experimental studies have found that when teachers are randomly assigned to touch some of their students and not others with friendly pats on the back, those students who receive the rewarding touch are nearly twice as likely to volunteer comments in class. When medical doctors are experimentally assigned to touch some patients but not others, those patients who are touched in a warm fashion estimate the visit with their doctor to be two times longer than those patients who go untouched. Students touched by librarians while checking out books indicated a much more favorable attitude toward that bastion of good undergraduate fun—the library—than students who were not casually touched by the librarian. Touch is the original contact high.

TOUCH TO THRIVE

In her excellent book *Touch*, Tiffany Field, the world's leading investigator of touch, recounts a moving story about an incident in a nursing home for the elderly. Two elderly residents, a man and a woman, were found missing during the dinner hour. They were soon discovered in a small utility closet, embracing one another like old friends. The staff quickly deemed them "sex offenders" and prevented them from further contact. Isolated, they withdrew from friends and families. Within weeks they were dead.

It is not a stretch beyond rigorous empirical evidence to claim that touch is essential to our physical and mental vitality. Some of the earliest systematic observations on this theme came from studies of orphanages, where only seventy-five years ago mortality rates for infants hovered between 50 and 75 percent. In one, run by a warm, friendly, affectionate German woman, the infants thrived. In another, where the children received no touch, the orphans were undernourished, sickly, and more likely to die. In a more systematic comparison, Renée Spitz assessed how well infants were doing at two orphanages, one in which female convicts served as mother surrogates, the other a foundling home. In both the infants were given food and clothing and kept clean. The foundlings had better access to medical services and were kept in a cleaner environment, but they were deprived of touch. They fared worse in terms of life expectancy and cognitive development.

More controlled studies have yielded comparably striking results showing how critical touch is to thriving. Tiffany Field has found that massages given to premature babies lead, on average, to a 47 percent increase in weight gain. In another study, thirty human infants were observed in the course of a painful heel lance procedure, in which the infants' heels were cut by medical doctors. Some of the infants were held by their mothers in whole-body, skin-to-skin contact. Others underwent the procedure while swaddled in a crib. The infants who were touched during the procedure cried 82 percent less than the comparison infants, they grimaced 65 percent less, and they had lower heart rate during the procedure.

Touch alters not only our stress-related physiology but the development of the underlying physiological systems that render the human stress response more labile and strong. Responses to stress are governed by two populations of neurons in the central nervous system. One population in the paraventricular nucleus of the hypothalamus projects to the anterior pituitary, which produces the adrenocorticotropin hormone (ACTH) that causes the release from the adrenal gland of the stress hormone glucocorticoid. The second population of neurons resides in the amygdala and projects to a region known as the locus ceruleus, which, when stimulated, leads to the release of noradrenaline. These respective neural populations ultimately stimulate the liver, the heart and circulatory system, and different organs. These organs then kick into action (for example, the liver increases glucose output to maintain stable blood sugar levels) to support stress-related behavior.

In stunning research, Darlene Francis and Michael Meaney have examined how a pattern of licking and grooming directed by rat mothers, or dams, to their pups alters the development of the HPA axis, which is the body's stress system (see figure below). The researchers first identified those rat mothers who provided a high level of licking and grooming to their pups. This was accomplished by observing patterns of licking, grooming, blanket postures where the mother lies over the pups, and various nursing postures. Francis and Meaney assessed this family of tactile behaviors

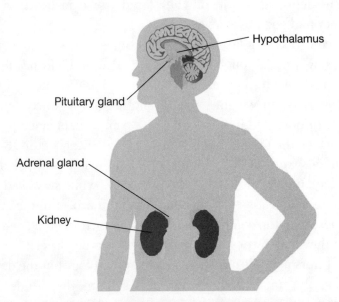

The hypothalamic-pituitary-adrenal (HPA) axis. The HPA axis is the body's stress system. Threatening events quickly activate the amygdala, which sends a signal to the hypothalamus, setting in motion a cascade of physiological events that prepare the body to fight or flee. The hypothalamus secretes corticotropin releasing factor (CRF), which stimulates the anterior pituitary gland, which in turn releases adrenocorticotropin hormone (ACTH) into the bloodstream. ACTH reaches the adrenal glands within a couple of minutes, which releases the stress hormone cortisol into the bloodstream, which stimulates the fight/flight response at different organs in the body.

five times a day for six to eight days postpartum. The licking and grooming are relatively infrequent, constituting about 10 percent of the interactions between mother and pup (the most common postpartum observation is no contact or arched-back nursing). There is tremendous variation in how much each mother licks and grooms her pups. The amount of touch provided has profound consequences.

Francis and Meaney have found that mothers who lick and groom a lot alter the HPA axes of their offspring. They produce young rats who are more resilient to stress. As adults, offspring of mothers who lick and groom a lot show reduced levels of ACTH and the stress hormone corticosterone in response to being stressfully restrained. They show reduced startle responses and a greater tendency to explore novel environments and foods. Perhaps most dramatically, they show reduced receptor levels of stress-related neurons in the brain (decreased corticotropin releasing factor receptors in the locus ceruleus; decreased central benzodiazepine receptors in the amygdala). Touch altered these animals' nervous systems. Early touch in a rat pup's life leads to a more resilient and calm rat later in life, endowed with a more robust immune system.

Of course, it is next to impossible to do this kind of precise study of touch and the HPA axis in young humans. A recent study, though, indicates that touch reduces our stress-related physiology. Jim Coan and Richie Davidson had participants wait for a painful burst of white noise—a source of stress—while resting in a fMRI scanner and having images of their brain taken. For the control participants, this stressful period of waiting triggered activation in the amygdala. These participants were showing a well-replicated brain reaction to threat. Other participants waited for the burst of white noise while their romantic partner touched their arm. These participants showed no amygdala response to the threat. Touch turned off the threat switch in the brain.

Touch is woven into our daily exchanges. Pats on the back, handshakes, hands resting on shoulders and arms, and playful nudges are barely noticed as we move through the day. Yet these touches alter others' nervous systems toward patterns of activation more conducive to higher *jen* ratios. The stroking of touch-sensitive neurons in the skin sends signals to one reward region of the brain—the orbitofrontal cortex, which activates release of oxytocin and endorphins. At the same time, pleasurable touch reduces activation of the HPA axis, the provenance of stress and anxiety. To touch, Michelangelo said, is "to give life."

TOUCH AND TRUST

Like many an American family, when our children were very young we had sleep arrangements that would leave a hunter-gatherer family, or their prim and proper Victorian counterpart, scratching their heads in disbelief. This was in part because we were torn between these two poles of sleep philosophy—the evolutionarily old and near-universal practice of family members sleeping in physical proximity to one another, and the Victorian innovation of making children sleep alone in dark rooms roiling with shadowy images of monsters and demons.

The product of such cultural ambivalence, we naturally arrived at elaborate bedtime rituals to get our two daughters to sleep. When they were quite young, say four and two, our bedtime ritual took an hour and involved the following: two and sometimes three fairy tales; two stories from my childhood, as long as I was younger than twelve and the stories involved some kind of mammal, some slapstick action on my part, and a subtle moral; one song for each daughter; and then patterns of sitting and lying next to each daughter. Of course one has only so many good stories to tell, and the best selection of the world's fairy tales can only be so enthralling to the jangled parent's imagination. Like many parents, I often was at wit's end during this ritual, plagued by visions of walking out the front door and hitchhiking across the country, and resorting, as a way to pass the time, to counting the minutes until their puberty would cast me out of their room.

I was saved by touch. Toward the end of the nightly ritual my younger daughter, Serafina, who entered the world with outstretched hand before the crown of her head, preferred that I sit next to her bed, which I did reliably, and eventually with anticipation. The reason: She would gently stroke the back of my hair near the neck as she fell asleep. She was targeting a region near the top of my spinal cord, where the vagus nerve, loaded with oxytocin receptors, originates and, I am convinced, is stimulated by such patterns of touch. We were engaging in a trade with ancient evolutionary origins. I offered my protective presence as she finally closed her eyes and drifted into the dreamy quiet of the dark. She offered me the most pleasurable of touches to the back of my neck, a kind of touch that was as potent a trigger of my pro-social nervous system—the orbitofrontal cortex, oxytocin (the little I have), the vagus nerve—as I have ever experienced.

The right touch—not some uncle squeezing your cheeks purple or a bully giving you a twist to the arm—creates trust and long-term cooperative exchange. Through its rewarding features, touch can be a glue of trading relations between kith and kin. One of the first to document this systematically was Frans de Waal, who has studied the role of touch in the patterns of food exchange in chimpanzees. Sure enough, chimpanzees use touch as a reward, and as a means of asking for favors. De Waal observed over 5,000 instances of food sharing in captive chimpanzees, carefully noting the patterns of who shared with whom in the troop. Chimpanzees, like our hominid predecessors, have a strong urge to share and to avoid hoarding. De Waal found that chimps were much more likely to share with those who shared previously with them and with chimps who had groomed them earlier in the day. They systematically traded calories for touch.

The same is true of humans: touching triggers trust and generosity. In one study, participants were asked to sign a petition in support of a particular issue of local importance. Those participants who were touched signed 81 percent of the time. Those who were not touched during the request volunteered to sign at a rate of 55 percent. In a recent study, Robert Kurzban put a participant into the prisoner's dilemma game, which gives participants the opportunity to compete or cooperate with a fellow player. As they were about to play the game, the experimenter lightly touched the participants on the back, creating an atmosphere of trust and generosity. This seemingly inconsequential act was enough to shift the frame of the game from one of competition to one of cooperation. Those participants who were touched were much more likely to cooperate.

It is not a coincidence that greeting rituals around the world systematically involve touch. Irenäus Eibl-Eibesfeldt has catalogued greeting rituals with surreptitious photography in remote cultures in Africa, Asia, Europe, New Guinea, and elsewhere. First contacts elicit, in ritualistic fashion, many of the tools that promote cooperation—submissive bows, smiles, open-handed gestures of cooperation. But they most systematically involve touch and skin-to-skin contact in various forms: handshakes, chest to chest embraces, and, in subtler forms than those used by rat dams and pups, varieties of kisses. Touching and trusting go hand in hand.

TOUCH AND THE SPREAD OF GOODNESS

If there is a consensus in the scientific study of morality and human goodness, it is that emotions like sympathy, love, and gratitude are the engines of everyday *jen*. For Charles Darwin, sympathy was a cardinal moral emotion.

Buoyed by this claim, ten years ago I began a search to document the nonverbal displays of sympathy and gratitude. Both emotions involve a powerful concern for enhancing the welfare of others, and a willingness to subordinate the demands of self-interest in the service of another. For cooperation to spread in groups, the contagious goodness hypothesis would suggest, sympathy and gratitude should possess reliable and evocative signals, allowing group members to readily discern the cooperative intent of others and, when feeling altruistically inclined, evoke cooperative tendencies in others.

Evidence of distinct nonverbal displays of sympathy and gratitude would then justify the search for the evolutionary origins of these emotions in other primates and mammals, and in our nervous system as well. So I began my quest for the signs of these emotions by turning to what I knew best—the face. I concentrated on sympathy, confident that I would document a unique facial display of

The facial actions of sympathy

this emotion. This work was based on Nancy Eisenberg's important finding that when people feel sympathy and are inclined to help others in need, they show a concerned eyebrow and pressed lip. When I presented images of this display to participants (see figure), and asked them to judge the emotion shown in the face, my hopes were dashed. Participants had only a faint idea what the person with the "sympathy face" was feeling. A few said compassion and sympathy, elevating my hopes. The majority, however, said things such as: she looked like she was concentrating or confused; still others volunteered answers like "she's drunk or stoned" or "she's constipated." Those states certainly did not offer evolutionary clues about this most virtuous of emotions, sympathy.

So like a good emotion researcher, I turned to the next best studied modality of emotional communication—the voice. Here Emiliana Simon-Thomas and I had twenty-two different individuals utter sounds that they would normally use to communicate a variety of different emotions, including sympathy, love, and gratitude. We achieved modest, but unremarkable, success: When I presented these vocalizations of sympathy, love, and gratitude to a pool of participants and asked them to judge the emotions in each voice, about 50 percent correctly identified the vocalizations of sympathy as communicating that emotion (see figure below). They had no idea, however, what to make of the vocalizations of love and gratitude. The most pro-social of the emotions did not seem to register in the face and voice.

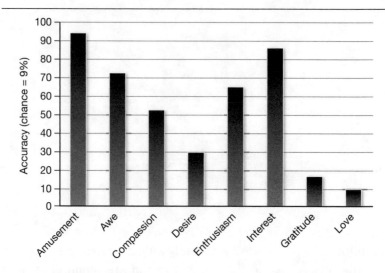

Accuracy rates in judging vocal bursts of emotion.

Thankfully, graduate students wander into my lab with interests I've never imagined. Matt Hertenstein, now a professor at DePauw University, suggested that we look at touch. Perhaps it is with touch that we convey these most pro-social emotions so critical to *jen* and the spread of goodness to others. Certainly studies of touch and the orbitofrontal cortex, oxytocin, reduced amygdala response, and reduced cortisol would suggest so. Perhaps William James was right in his observation that "Touch is both the alpha and omega of affection." So Matt and I designed an experiment motivated by a simple question: Can we communicate sympathy, love, and gratitude through touch?

Clearly, the more general requirements of the study were straightforward—one person, the toucher, would be given the task of communicating sympathy, love, and gratitude and other emotions to another person—the touchee. The touchee would only be able to rely on tactile information in discerning the emotion conveyed in each touch. Our first version of the study was a disaster. In this version, our touchee sat blindfolded with earplugs in a lab room. The toucher was given a list of twelve emotions, including sympathy, gratitude, and love, and asked to touch the blindfolded individual, in any fashion within reason, to communicate these emotions. The touchee, who sat in a state of sensory deprivation, knew of the list of twelve emotions that were soon to descend upon his or her skin, and had the task of picking a term that best matched the touch that was just delivered.

The study more resembled a piece of performance art than science. One set of participants acting as the touchee found it to be a form of torture, sitting silently in a sightless and soundless world, ready to be poked in anger or soothingly stroked in compassion. Another portion of students, usually males, found the study to be exhilarating. I have the strong sense they would have paid good money to sit blindfolded and have female participants touch them to communicate love and gratitude.

So we turned to a primitive technology. We built a large barrier in a lab room, a wall to separate toucher and touchee. Part of this barrier included an opaque black curtain. The curtain prevented any kind of communication between toucher and touchee—visual, auditory, olfactory—other than touch. First, both toucher and touchee reviewed the list of twelve emotions: anger, disgust, embarrassment, envy, fear, happiness, pride, sadness, surprise, and the three of interest—sympathy, love, and gratitude. The touchee bravely put his or her arm through the curtain and awaited twelve different touches, randomly ordered. For each touch, the touchee guessed which emotion was being communicated. The toucher could only make contact with the touchee's arm from elbow to hand to signal each emotion, using any form of touch. The touchee could not see any part of the touch because his or her arm was positioned on the toucher's side of the curtain.

Our measure of interest, represented in the table below, was the proportion of participants selecting the appropriate term to label the touch. As you can see, people can reliably communicate well-studied emotions such as anger, disgust, or fear with a one- or two-second touch of another's forearm. Quite astonishing, really, was how well strangers could communicate sympathy, love, and gratitude with one-second touches to a stranger's forearm. Just as interesting were the emotions that our participants could not readily communicate with touch, such as embarrassment and pride, which are founded upon a sense of how others regard the self.

PRIMARY CHOICE		SECONDARY CHOICE	
WELL-STUDIED EMOTIONS			
ANGER	57	DISGUST	15
DISGUST	63	ANGER	10
FEAR	51	ANGER	14
SADNESS	16	SYMPATHY	35
SURPRISE	24	FEAR	17
HAPPINESS	30	GRATITUDE	21
SELF-CONSCIOUS EMOTIONS			
EMBARRASSMENT	18	DISGUST	16
ENVY	21	DISGUST	12
PRIDE	18	GRATITUDE	25
PRO-SOCIAL EMOTIONS			
LOVE	51	SYMPATHY	28
GRATITUDE	55	SYMPATHY	16
SYMPATHY	57	LOVE	17

Humans can communicate emotion with one-second touches to the forearm.

We replicated this study in Spain, known as a high-touch culture, and, our participants were a bit better able to decode emotions through touch.

Our study also involved all possible gender combinations—women touching women and men, and men touching women and men. Here we found two gender differences that speak volumes about the different planets women and men are claimed to originate from. The female participants' attempts to communicate anger via touch to the male touchees were a failure. The male participants had no idea what the females were doing, and the males' judgment data amounted to a random collection of guesses at what the women were trying to convey. A woman's anger does not seem to penetrate the skin of a man. Regrettably, it gets worse. The male participants' attempts to communicate sympathy to the females were absolutely unintelligible to the females; the males' attempts at sympathy fell on deaf skin, so to speak.

When we coded what people were doing when touching to communicate the different emotions, we documented behavior that traces back in evolutionary time to our hominid predecessors. Sympathy was conveyed most regularly with a soothing, slow stroke to the arm, no doubt designed to trigger maximal activation in those Merkel cells in the epidermis, generating neural impulses directed toward compassion regions of the brain and nervous system. Gratitude, very interestingly, was reliably signaled in a firm clasp of the forearm, adorned with a slight but clear shake of reassurance.

Sympathy and gratitude are central players in the social contract, motivating actions in the service of others. These are not recent arrivals in evolutionary history or contrivances of a particular culture. They are emotions that are embodied in tactile exchanges that have been honed by thousands of generations of hominid evolution, so that today, with a simple touch to the forearm, the receiver of the touch can discern sympathy from gratitude from love.

HOOPS AND PEDICURES

Five minutes at the chimpanzee compound at your local zoo will reveal how pervasive touch is. You'll see mothers grooming their babies, alpha males picking at the hair of close competitors; two cavorting juveniles, ricocheting around the branches, suddenly stop their antics to groom. In fact, primatologists estimate that chimpanzees, our closest primate relatives, devote upwards of 20 percent of their waking hours to grooming. Grooming is so vital to the primate slow loris, *Nycticebus tardigradus*, that this species has evolved a single nail, known as the toilet claw (in the etymology of

 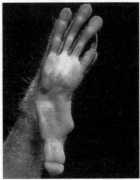

The slow loris and its toilet claw (the short, curved digit).

Chimpanzees grooming

toilet, "toilette" came first, and referred to a room to groom in), which evolved to enable frequent grooming.

The first interpretation of the prevalence of grooming in primates, sound and intuitive, was that they were simply ridding one another of parasites, thus enhancing the chances of physical survival. No doubt the need to get rid of bacteria and virus-infested parasites got primates grooming in the first place. Observant primatologists, however, were quick to document episodes of grooming that did not fit the parasite thesis. Primates groom to play, to reconcile, to soothe, to get close, and prior to copulation, with no visible intention of finding parasites. More convincingly, primates groom regularly when there are no known parasites in the physical environment.

This led Robin Dunbar to observe that perhaps grooming is like human gossip. Grooming is a casual exchange of daily living that bonds individuals to one another. It is a glue of our social relations. And so it is with human touch: Touch spreads goodwill, cooperation, and trust.

We live in a touch-deprived culture. The impoverishment of touch in U.S. society owes its deep roots to the Puritans, well known for their attempts at extirpating ordinary human delights— dance, laughter, theatrical drama, and touch. A finger could be pointed at an obvious target, repressive Victorian culture. In the upper-class stratum of Edith Wharton, infant was separated from the breast of mother, sleeping children from sleeping parents, dreaming wives from dreaming husbands, and skin was covered to remain inaccessible to the human hand. A product of this cultural legacy, the influential psychologist and educator John Watson observed: "There is a sensible way of treating children. Never hug and kiss them, never let them sit in your lap. If you must, kiss them once on the forehead when they say good night and shake hands with them in the morning."

Today, the signs of touch deprivation are abundant. Teachers are actively prevented from giving students pats on the back or, God forbid, a hug, out of a fear of allegations of sexual harassment (I'd bet my life savings that any teacher worth his or her salt knows the right kind of touch to encourage students). Parenting manuals discourage too much physical contact on the assumption that the child might grow up to be "overenmeshed." In a recent observational study of the frequency of touch in cafés in different parts of the world, University of Florida psychologist S. M. Jourard observed two people in conversation over a cup of coffee. In London, not a single touch was observed; in Florida, 2; in Paris, 110; and in San Juan, Puerto Rico, 180.

Yet the instinct to touch is too wired into us to remove from our daily affairs. The ancient, evolved tendency to touch has obvious cultural translations—massages. The instinct to touch is evident in quirkier cultural forms—the hundreds of cuddle clubs that have sprung up across the country, where people lie around in euphoric, sleepy-eyed piles, cuddling nonsexually (so they claim). The need to touch is hidden in various cultural forms: manicures, pedicures, haircuts, and, I would wager, a rate of visits to medical doctors that would startle insurance companies. This instinct for touch is the source of economic innovation, like the soft carriers that allow parents to carry children where they want to be—close to the front of the body. Compared to infants carried in harder infant seats, infants who were carried in soft infant carriers that put them in close physical contact with their parents are more securely attached to their parents and more willing to explore novel environments. Thanks to Tiffany Field, touch has been integrated into medical treatment. There have now been over ninety scientific studies of touch therapy, and these studies have found that regular touch helps premature infants (who used to be deprived of physical contact), depressed teenage mothers, the elderly in nursing homes, children with autism, ADHD boys, children suffering from asthma and diabetes, and people suffering from disease.

The ancient language of touch is a backbone of cooperation; it is a source of high *jen* ratios. For the past twenty-five years I have played pickup basketball twice a week, participating in the most democratic institution in the United States. I have played with people from all walks of life—Andover grads, kids from the projects in Brockton, Massachusetts, novelists, medical doctors, seventy-year-olds, lapsed drug dealers, lipstick lesbians, yoga instructors, music producers, chefs, psychotherapists, tattooed firemen, cops, performance artists, and drifters off the street playing in paper-thin shoes. Points are scored. Winners win and keep the court. Losers lose and line up for the next game. Calls are made and contested, especially when the game is on the line. Ten bodies, each weighing on average 200 pounds, crash into each other for hours at a time, with a force that sprains ankles, breaks noses, blackens eyes, and wears down the knee cartilage until it's bone on bone in the middle of life.

I estimate that I've played approximately 4,500 games, from Brockton to Pau, France to Haight-Ashbury in San Francisco. Those games have involved my oldest friends and people I'll never see again. And in those 4,500 games, where loud voices and thrown elbows reign, I have never once seen a fight break out. Sure, there are dramatic confrontations, and many a shove under the boards. But I've never seen a punch thrown or anything remotely resembling unadulterated aggression. That level of violence (0) proves pickup basketball to be more peaceful than randomly sampled interactions between marital partners, siblings, family members at

Thanksgiving, crowds celebrating their football team's triumph, people parking to go to the theater. At the end of the game, there is most typically laughter, respect, and a faith in the human project. The rest of the day is more peaceful.

Why? Because the violent physicality of basketball is transformed by touch. Teammates bump fists when the game begins. During the game opponents lean into each other, hand check to the hips, push forearms to the back and chest. Defenders bear-hug to stop a drive down the lane. Opponents slap rumps at a good play. There are high fives at the game's end. The visible physics of basketball is incommensurably violent—bodies colliding at near-full speed. The language of touch in the pickup game neutralizes the aggressive intent of these actions.

How so? I asked my daughters about their first experience of another cultural form that revolves around touch—a pedicure. They had just luxuriated in one as a special treat with their mother. Here is their response:

NATALIE: It felt like comfort.

DAD: Why?

NATALIE: Because they massaged your leg. You sit in a chair and they massage you.

SERAFINA: And they put pretty nail polish on perfectly. Except it stinks.

DAD: So what did it feel like?

NATALIE: It was kind of painful, the scraping of your nails. But the leg massage felt like vibrations in your back, like someone was humming.

The humming vibration in the back is how touch has evolved to spread goodness and shift people's *jen* ratios toward higher, loftier values. Touch has been made more fragile by cultural forces that prevent people from coming into contact with one another. The leg massage (the real purpose of the pedicure) and all of our touch rituals (pickup basketball, haircuts, handshakes, rough-and-tumble play, pats on the back) trigger activation in the orbitofrontal cortex and the release of opioids and oxytocin. They trigger the activation of the vagus nerve, the nerve bundle in the body devoted to trust and social connection, which, when activated, indeed feels like humming vibrations in your back. And if we had precise enough measures, we probably would find that that incidental leg massage, not described in the ad for the pedicure, shifted the stress regions of Natalie's nervous system—HPA axis activity—to more peaceful settings, and amplified her physiology of trust and goodwill, perhaps, one might hope, in a permanent way.

10

Love

O N   A   C O L D February weekend, my wife, Mollie, and I and our daughters, then 7 and 5, made the two-hour trip to Año Nuevo State Park, near Monterey, California. Our aim was to weather the winter storms to view the natural spectacle of migrating elephant seals on their way from Baja to Alaska. We were going in the spirit of Charles Darwin, seeking to study the social patterns of other species to glean insights about our own.

Gale-force winds prevailed. Whipping sheets of sand prickled the young families who gamely trudged on through swirling sand dunes. Failing to appreciate the youthful attention spans of half her audience, our park ranger guide droned on about alpha males, harems, mating rituals, ululations, gestation cycles, and migratory patterns. As we marched on, heads down, eyes shielded by hoodies, children burst into tears, soothing lollipops—last acts of desperate bribery—dropped into the sand.

At last we arrived at a little hill, a purchase, where we were to lie quietly to watch the beached elephant seals gathered below. We lay prostrate on the cool sand in a layer of warm air below the gusts, steadying our binoculars and cameras upon the elephant seals. The enormous 4,500-pound alpha male, heavier than the average SUV, guarded dozens of the females in his harem, each roughly a fourth his size. Occasionally the alpha male would galumph over to a female and flop on top of her. She was lost to view under the rippling gyrations of his fat. At the sight of this burst of passion, other males, poised on the periphery of the harem, would make their move, flopping toward nearby females. Such an intrusion proved more attention-worthy to the alpha male than the paramour below; in the only alacrity he was capable of, he would charge, blubber rolling, toward the intruder. Within ten feet of contact the alpha male would rise up, with weird trunklike snout, and ululate as loud as a corn thrasher. This pattern of rest, attempt at copulation, intrusion, and confrontation went on endlessly. No cuddling, no play, no frolicking, no snout-to-snout nuzzling or mutual gaze in sight.

Our guide rounded us up and led us down a path to the rolling Pacific Ocean to see whether we might spot baby elephant seals, born just a couple of months ago in Baja. Tiny elephant seals at play in the surf would rescue the falling spirits of my daughters. Instead, near the final dune we were to climb, off to the side of our driftwood-marked path, lay a dead baby elephant seal. Our guide explained: "On occasion the male elephant seal, following an ancient evolutionary instinct, will accidentally try to mate with a baby, often to dire ends." For the rest of the tour, my daughters clung to me, heads buried in my shoulders.

At the end of the tour, after a few polite questions, tips, and half-hearted "thank yous," we returned to our overstuffed Subaru. Natalie and Serafina sat in their car seats, solemn and quiet. I could feel them trying to map words their parents had used just prior to this misadventure ("family," "kind of like husbands and wives," "new babies," "love") onto the raw spectacle of elephant seal reproduction. What a flawed endeavor it is to map a concept in the English language ("love," "family," "husbands and wives") onto the immense variety of reproductive arrangements in nature or, for that matter, the complexities and nuances of love.

Had I had the right words and temerity (and had they been several years older), I would have reassured my daughters with new studies from the evolutionary biology of reproduction, so sharply summarized in *The Red Queen* by Matt Ridley and *The Ant and the Peacock* by Helena Cronin. Elephant seals are a tournament species, where males devote much of their energy and psyche to violent, winner-take-all competitions for large harems. Humans are more on the pair-bonding end of the continuum, closer to the gibbon, the delicate sea horse, certain voles, and many bird species. In the 8,000 pair-bonding species, like humans, the male is less differentiated than the female in terms of size or florid color, and there is less radical variability in the reproductive outcomes of males (in elephant seals such as those that we observed, almost all offspring are sired by the alpha male). Their future boyfriends, still many years away, would not hoard dozens of girlfriends in the lunchroom and fill entire daycares with their offspring. Instead, they would be theirs and only theirs, at least for awhile.

There is more. In humans, the default is for monogamy, and not the harem tendencies of the elephant seals (no need to unsettle their trust in my marriage with a discussion of the universality of serial monogamy). Sure, one can find elephant seal-like arrangements in human history, in particular in the early emergence of civilizations around the world, when powerful kings started to hoard resources and claim harems in the thousands. The Inca sun king Atahualpa kept 1,500 women in "houses of virgins" located throughout his kingdom, chosen for their pristine beauty most typically before the age of eight. The Indian emperor Udayama kept 16,000 consorts in apartments ringed by fire and guarded by eunuchs. But in early hunter-gatherer culture and in contemporary industrialized cultures, the robust tendency is toward serial monogamy and the intricate challenges of one woman and one man conducting a life together.

Unlike those elephant seals, I would have continued, human males actively contribute to the raising of the offspring. In over 90 percent of mammals, the female is the sole provider of care to offspring; the male doesn't lift a hand or change a metaphorical diaper. We are different. Human males have the capacity for levels of care for offspring reminiscent of the devotion of the sea horse, the gibbon, and many birds. Tens of thousands of fathers in the United States are primary caretakers, changing diapers, pushing swings, reading *Babar* and the tongue-twisting wisdom of Dr. Seuss, negotiating sibling conflicts, playing rough-and-tumble, speaking "motherese."

I would have reminded my daughters that we make friendships. Humans in nonreproductive relations do not flop around like those elephant seals, little cognizant of one another, except in confrontations over mating opportunities. Humans feel deep love for nonkin, in particular for friends. This they would have readily grasped, for they already had folded into devoted friendships. We even feel elevating love for our own kind, humanity, and other species.

Had I had the words or notion, I would have told my daughters that outside of the love they feel for each other (and other kin), there are four great loves in life. There is the love between parent and child, the passion for sexual partners, the enduring devotion for long-term pair-bonders, and the softer but rock-solid love for nonkin, most typically friends and fellow humans.

THE NOT-SO-INVISIBLE HAND

On a chilly January morning in 1800, a dirty, naked twelve-year-old boy, scampering around on hands and feet, was spotted digging for potatoes in the fields of the French village of Saint-Sernin. He was an abandoned child, not uncommon during the era. He had survived for years on his own in the forest, scavenging for acorns and hunting small animals, deprived of the warm care of parents.

The owner of the field captured the wild-eyed boy and took him home. The boy, soon to be named Victor, prowled restlessly on all fours. He refused to wear clothes. He defecated in public and rejected all food except acorns and potatoes. His communication was restricted to grunts, howls, and cackles. He was unresponsive to the human voice and language but would turn quickly at the sounds of nuts being cracked. He never smiled, cried, touched, or met the gaze of other humans.

Eventually, Jean Itard, a twenty-six-year-old doctor from the Paris Deaf-Mute Institute, took Victor—the "Wild Boy of Aveyron"—into his custody, and devoted five years to teaching Victor language and the intricacies of human ways. There were telling successes: Victor did learn to wear clothes, sleep in beds, eat at a table, and take baths. Most notably, he came to feel affection for Jean Itard.

There were telling failures. In spite of the intensive instruction, Victor only learned a few words. He never learned to get along with others (at a dinner party at a wealthy socialite's home, designed to show off his progress, he wolfed down his food, stuffed desserts into his pockets, stripped to his underwear, and leaped through the trees like a monkey). Victor resembled the other thirty-five documented cases of feral children: They do not develop language, morals, or manners; they remain largely unresponsive to humans; they fail to fold into cooperative relations with other people; they show no sexual interest; and they lack self-awareness. The first great love is what Victor never felt, that between parent or caretaker and child. This love enables what it means to be human; it turns on our tendency toward *jen*.

Philosophers (to some extent), poets (to a greater extent), and novelists (to an even greater extent) have long recognized that the love between parent and child is the foundation of human mind, character, and culture. It would take a maverick intellectual, John Bowlby, integrating the latest in evolutionist thinking and the musings of Freud, to spur the scientific study of parent-child love. Given the profound vulnerability of human offspring, Bowlby theorized, evolution has designed an "attachment system": biologically based patterns of behavior and feeling that bind caretaker and vulnerable infant to one another, in devoted, skin-to-skin, voice-to-voice, eye-to-eye contact. When Bowlby's collaborator Mary Ainsworth did early observational research in Uganda on the attachment behaviors of young infants there, she documented familial universals: only in the presence of their mothers, Ugandan infants showed specific kinds of crying, smiling, and endearing vocalizations, clapping and lifting the arms when the mother approached, burying the face in the mother's lap, hugging, kissing, and clinging to the mother, and distress vocalizations when the mother moved away. Just as reliable are the attachment behaviors of caretakers: skin-to-skin, chest-to-chest contact, cradling, massaging touch, playful coos and sighs, eye contact, "motherese," soft-toned songs at night, joint smiling and antiphonal laughter.

Mammals just aren't mammals when deprived of the love between caretaker and offspring. In Harry Harlow's well-known research, rhesus monkeys raised in isolation, deprived of contact with parents (and peers), grew up to be the wild boys of Aveyron of their group: profoundly fearful, inept in forming relationships with peers, as likely to attack potential sexual partners as court them; attempting to copulate with same-sex peers. Elephants in some areas of Africa develop without their loving parents, who have been slaughtered for the ivory in their tusks. These adolescent elephants show pathological forms of aggression, looking like the worst of our sociopaths, killing rhinoceroses for sport, for example.

These early attachment experiences, dozens of human studies show, lay the foundation of the capacity to connect. In the thinking of John Bowlby, these early experiences of love alter our *jen* ratios or, in Bowlby's terminology, the individual's "working model" of intimacy, trust, and the goodness of others, deep, early beliefs that shape our peer relations, work dynamics, ensuing adventures in our own families, and engagement in communities. Individuals who report a secure attachment style feel comfortable with intimacy and desire to be close to others during times of threat and uncertainty. They were likely raised by parents who were responsive to their early needs and emotions. And as adults, these individuals enjoy healthily high *jen* ratios. People who report a sense of secure attachment perceive their partners to be a steady source of support and love. They look charitably upon their partner's criticism, tension, and insensitivity, putting a positive spin upon these struggles of intimate life. And as life progresses, securely attached individuals feel greater satisfaction in their current romantic relationships, they are about half as likely to divorce as other individuals, and they consistently report a greater sense of contentment and meaning in life.

Anxiously attached individuals, by contrast, feel a deep sense of uncertainty about their attachment to others; they feel that others do not give enough and are not reliable sources of intimacy and love. Their parents, research shows, were less responsive and warm and more tense, anxious, and distant in their minute-by-minute interactions. A quick study of a morning in such a house would find a more impoverished vocabulary of attachment behaviors—encouraging touch, warm smiles, brief eye contact, and playful vocalizations—and more sighs of exasperation, remote gazes, and painful touch. These more anxiously attached individuals have greater difficulties in their subsequent bonds—greater dissatisfaction, cynicism, distrust, and criticism. These tendencies suffuse every moment of their intimate relations. When Chris Fraley and Phil Shaver surreptitiously observed romantic partners as they said good-bye in airports, anxiously attached individuals expressed great fear and sadness as their partners headed down the walkway, privately suspecting that this would be the last they would see of their beloved. Anxiously attached individuals are more likely to interpret life events in pessimistic, threatening fashion, which increases the chances of depression. They are more likely to suffer from eating disorders, maladaptive drinking, and substance abuse, in part to reduce their distress and anxiety. They are more likely to have intimate relationships that dissolve in bitterness.

The first great love of life begins upon leaving the womb. It lasts, in the words of John Bowlby, "from cradle to grave." It is laid down in a rich vocabulary of touch, voice, gaze, and facial display, it is evident in the merging of minds, heartbeats, and nervous systems of caretaker and young child. These processes establish deep patterns of neural response in the pro-social nervous system—growth in tactile receptors in the skin, strengthening of the oxytocin system (which is damaged in orphans), the setting of the HPA axis to less stressful levels, lighting up of reward centers in the brain. These early attachment experiences are laid down so early we can't consciously remember them, for the regions of the brain involved in memory—the hippocampus in particular—aren't fully functioning until age two or so. But they are felt every moment of life, in the trust of a stranger, in the willingness to speak out and fail, in the devotion to a romantic partner in times of difficulty, in the sense of hope, and in the devotion one feels for one's own children. If it goes well, that early love is felt as the encouraging, not-so-invisible warm hand on your back as you move through life.

THE ELEMENTS OF DESIRE

A *lek* (Swedish for "play") is the singles bar for many bird species. It is a small patch of ground where the males of a species congregate and set up shop to take their shot at seducing their female counterparts. The male bowerbird, for example, will build elaborate bowers of sticks, leaves, bottle caps, and hot commodity items like the bird of paradise's feathers to show off his resource-acquiring abilities (see photograph). Like young women arriving at the dance after a trip to the bathroom, female bowerbirds arrive at a set time at the leks, inspect each male, engage in a few courtship head bobs and coos, and then converge on a couple of males who seem most worthy (that is, resource-rich) to mate with.

It does not take a great leap of the imagination to recognize human leks—junior proms and Sadie Hawkins dances, bars, nightclubs, Bible study groups, coffee and copy machines at the office, singles hikers in the Sierras, activist meetings—where desire is negotiated according to our own ancient patterns of courtship. This ancient language of desire catapults us, heads spinning, into reproductive relations. Before cataloguing this language of desire, it is worth considering two underappreciated qualities of human desire that might be taken for granted. The first is that human desire channels us into monogamous bonds. This is not the trajectory of desire in our closest primate relatives. In gorillas, resource-rich alpha males lord over harems, while other males do their best to sneak in surrep-

The male bowerbird prepares for seduction.

titious copulations—like those elephant seal beta males. In chimpanzees, all is quiet on the sexual front until the female goes into
estrus. At this time she most typically mates indiscriminately with
dozens of males each day, often requiring up to 3,000 copulations
prior to pregnancy. And bonobos wage an all-out, polyamorous
Haight-Ashbury lovefest, using sex for just about every purpose: to
reproduce, to form friendships, to help share food, to play, to pass
the time.

Putting aside your bonobo envy, it is important to appreciate that
human desire, at least in the moment, is singular. It is oriented
toward one person; it pair-bonds. The most obvious reason for this
is that our big-brained, ultravulnerable offspring required multiple
caretakers, including fathers. Another factor, suggests Matt Ridley
in *The Red Queen*, is our love of meat. Some 1.6 million years ago,
our foraging, group-dwelling hominid predecessors started eating
meat. The provision of meat is a probabilistic affair, and bound
males into dependent trade relations. This focus of early hominid
dietary activity prevented any single male from hoarding all the
resources—a precondition for harems—and kept early hominids in
pair-bonding relations.

If you need further proof of our pair-bonding predilections, just
look at a few males' testicles. In species with polygamous females,
males have outsized testicles that produce copious amounts of
sperm to win in the game of sperm competition with other males.
Thus, in chimps, with their promiscuous females, the male's testicles on average are two times larger than those of the gorilla, whose
females mate in serial and monogamous fashion with one alpha
male. In the right whale, whose females are polygamous, the testicles of the male weigh half a ton, or 1 percent of its body weight
(two pounds on a 200-pound human). The right whale's testicles
greatly outweigh those of the male in the pair-bonding gray whale.
Human testicle size reveals us to be more on the pair-bonding end
of the continuum. Sexual desire is the rocket booster that moves us
toward that arrangement.

Human desire is just as remarkable in that it leads to sex and intimacy unrelated to procreation. Long before the birth control pill
revolutionized intimate life by freeing sexual behavior from repro-

ductive outcome, the same was happening in human evolution. Females of our closest primate relatives advertise their reproductive readiness with swollen, colorful sexual regions—displays that shock and astonish in the primate section of your local zoo. Human females, in contrast, have evolved concealed ovulation. As a result, women and men do not necessarily know whether their desire will lead to reproductive outcome (although a woman is more likely to initiate sex, masturbate, have affairs, and be accompanied by her husband during ovulation; and pole dancers earn bigger tips, Geoffrey Miller has recently found, at the peak of their ovulation). Concealed ovulation evolved, we now know, to prevent stepfather infanticide, which is unnervingly common in mammals, and seen in many rodent species, lions, and many primates. Concealed ovulation keeps males guessing about whether offspring are theirs, thus reducing the likelihood of infanticide. Concealed ovulation also allows women and men to have sex throughout the female's cycle— an ongoing incentive for the male to remain in a relationship and contribute to the raising of such resource-dependent, vulnerable offspring.

The specific language of desire, which propels potential partners toward one another, has been documented by Givens and Perper. These researchers spent hundreds of hours hiding behind ferns and jukeboxes, laboriously documenting four-or five-second bursts of nonverbal behavior amid the lambent light and Lionel Richie tunes of 1980s singles bars. They homed in on those microscopic behaviors that predict whether women and men will pursue a romantic encounter—a shared drink, an exchange of phone numbers, leaving the bar with buoyant step, arm in arm.

In the initial attention-getting phase, women walk with arched back and swaying hips, amplifying the extent to which their bodies take on that platonic form of beauty—the hourglass figure. Women resort to the well-known universal—the hair flip—which dominates the field of vision of the male of interest, who is nonchalantly sipping his third Bud. Women (and men) smile coyly, lips puckered, head turned away, but eyes dropping in to make eye contact for a millisecond or two (see figure below).

Men counter with behaviors that amplify their physical size and assumed resource-holding potential. They rock back and forth on their heels and roll their shoulders. They raise their arms with exaggerated gestures, in ordering a round of drinks or stretching out, to show off their well-developed arms, the broad expanse of their shoulders, or expensive watches or prep-school pinkie rings. These brief signals honor time-honored principles in the game of sexual selection. The woman is drawing attention to her curves, fine skin, and full lips—signs of her sexual readiness and reproductive potential. The man is signaling that he has stature, resources, and good genes, appealing to the woman whose desire is conditioned by an awareness of the enormous costs of pregnancy, childbirth, and breastfeeding, which would be offset by a man of means and justified by a man with good genes.

Like courtship, momentary flirtation progresses toward more intimate phases. In the recognition phase, women and men gaze intently at each other; they express interest with raised eyebrows, singsong voice, melodious, voiced laughter, and subtle lip puckers. They turn to the exquisite language of touch, and all those receptors under the surface of the skin, to explore their interest in each other with provocative brushes of the arm, pats on the shoulder, or not-so-accidental bumps against one another that safely occur in the aftermath of a joke or in a pleasant, joshing-around tease. A slight touch to the shoulder that is ever so slightly firmer and more enduring than a polite pat reveals a desire beyond the typical exchange between friends or new acquaintances.

The coy smile, which is quickly detected as a sign of desire in our studies.

If all of this proceeds well, the potential partners move to the keeping-time phase. They begin to mirror each other's glances, laughter, gaze, gesture, and posture, as they share jokes, order drinks, disclose embarrassing snippets of the past, and search for commonalities. This kind of behavioral synchrony creates a sense of similarity, trust, and merging of self and other. In Plato's view, the two souls, having separated at birth but now reunited, are forming the perfect union with one another.

In many species, courtship behaviors stimulate the biology of reproduction. For the tree-dwelling African dove, flirtatious coos and head bows trigger the release of estrogen and luteinizing hormone in the female, and eventually ovulation. A stag's roar stimulates females to go into heat faster. The lowly snail shoots darts into potential sexual partners, which activates the snail's sexual organs (I dare not describe them). In humans this language of passion stimulates the experience of desire. In the throes of this kind of love, people experience an entirely different sense of time and a disarming loss of personal control and agency. A metaphorical switch in the mind is turned on (and the voice of cost-benefit, conventional rationality is turned off). People feel blown away, swept off their feet, knocked out, ill, feverish, and mad. They may eat and bathe less, stop seeing friends, neglect their homework and bills. The old definitions of the self are turned off, to make way for the establishment of an entirely new identity, one that emerges in the early delirium and upheaval of pair-bonding relationship and which will rearrange their lives.

This language of desire carries the couple toward a different kind of consummation than that observed in other species. The couple will likely make love face-to-face, unusual in the primate world. They will have sex in private. And alongside desire, our research finds, they will feel a deep sense of anxiety. The woman will wonder whether her new partner resembles the male caricatured all too readily in scientific research, the male eager to pursue short-term sexual strategies (one-night stands) to dispense with his daily production of 200 million sperm (in one study, 75 percent of college males were willing to go home with a female experimenter they had just met while walking on campus, and who had asked whether they were interested in a quickie). The man will feel his own anxiety, perhaps sensing that he is unlike any other primate in the degree to which he will be expected to sacrifice, to forgo other reproductive opportunities, and devote resources to his offspring, whom he will, again unlike any other primate, recognize as his own. They await the warm surround of romantic love to shut down these anxieties.

OPEN ARMS AND MOLECULES OF MONOGAMY

Each year 2.3 million couples wed in the United States. The average cost of a wedding is $20,000, which exceeds the average life savings of any American you might pick off the street. Guests lists are negotiated, dresses fitted, invitations embossed and mailed, appetizers and music selected. What follows is a surreal day of rapturous emotion, fathers crying, mothers spilling their glasses of wine, exlovers smoldering, recalled verse, besotted smooches, best friends in arms, and dancing children.

The wedding ceremony could rightfully be thought of as the most elaborate, expensive ritual in human history to fail. Approximately 47 percent of those individuals who stand at the altar, suffused with lofty emotions, uttering vows in hallowed words of devotion, will divorce, and they'll often go down in flames of hatred and litigation. Very often they'll divorce within a year or two of the ceremony, giving each other the finger, as did the divorcing parents of a friend of my parents', in the county courthouse's courtyard, or uttering "Tu es mort" as they sign the papers.

Or, you could think of the wedding ceremony as an astonishing success. Half of marriages make it. In spite of frequent surges of youthful desires and the mundane complexities of marriage, estimates of adultery suggest that only 11 to 20 percent of married partners have extramarital affairs. Compare that success rate in taming nonmonogamous sexual impulses with recent studies of abstinence programs provided to middle- and high-school students. These expensive, sophisticated products engineered by well-meaning social scientists fail abominably, and often lead teenagers to be more inclined to have sex or unsafe sex after such indoctrination.

In terms of its outcome, the wedding ceremony can be seen as a glass half empty or a glass half full, an interpretation that no doubt is shaped by our own experiences with the person we enjoyed that day with. In terms of its function, there is no doubt about the interpretation of why we go to such lengths in the wedding ceremony: It is a ritualized solution to the commitment problem. The wedding ceremony is our attempt as a culture to get two young partners to remain faithful to one another (and devoted to their offspring) in the face of so many compelling alternatives; to sacrifice their pursuit of sexual desire to the interests of their bond and their offspring. Culture's answer is to empty the bank account, bring every person you cherish into a sublimely beautiful locale, make public avowals, give expensive rings to one another, photograph every instant of the day in the event that memory fades, and head off into the sunset. Evolution's answer to the commitment problem is that emotion most favored by poets and rock stars alike: romantic love.

Romantic love enables the human mind to countervail self-interest. In the depths of romantic love, we idealize our partners; they take on unique, mythic qualities; we turn to deistic metaphors to describe our beloved. When Sandra Murray and her colleagues asked romantic partners to rate themselves and their partners in terms of different virtues (understanding, patient), positive traits (humorous, playful) and faults (plaintive, distant), they found that happier couples idealized their partners; they overestimated their partners' virtues (compared to the partners' self-descriptions) and underestimated their faults. In other studies, Murray and colleagues asked people to write about their partners' greatest fault—the source of endless vitriol in therapy sessions and divorce proceedings. Happier romantic partners were more likely to see virtues in faults and more likely to offer "yes, but" refutations of faults. A happier married wife would look at her lethargic husband on the couch, snoozing with the remote pressed into his cheek, and think, "yeah, but at least he is around more in the home and not cavorting at the sports bar or at the golf course all day Saturday."

Studies point to a neurological basis for romantic love's rose-colored glasses. Not too surprisingly, long-term committed romantic love is associated with activation in reward centers in the brain—the ventral anterior cingulate, the medial insula, the caudate and the putamen. More dramatically, romantic love deactivates threat detection regions of the brain—the right prefrontal cortical regions and the amygdala. The person in the throes of romantic love may be physiologically incapable of seeing all that is worrisome, problematic, or worthy of a skeptical second look.

And studies have begun to document the very chemical that promotes long-term devotion. We can pin our hopes on oxytocin, a mammalian hormone, or neuropeptide, consisting of nine amino acids, and involved in humans, as any midwife will tell you, in uterine contractions, milk letdown, and breastfeeding. Oxytocin is produced in the hypothalamus, an old region of the brain that coordinates basic behaviors related to food intake, reproduction, defense, and attack. It is then released into both brain and bloodstream, which is why it is called a neuropeptide. Receptors in the olfactory system, neural pathways associated with touch, and regions of the spinal cord that regulate the autonomic nervous system, especially the parasympathetic branch, including the vagus nerve (see chapter 11), await the chemical's arrival.

By activating touch and a calmer physiological state, oxytocin enables monogamous pair-bonding. This remarkable discovery emerged in Sue Carter and Tom Insel's comparisons of two nearly genetically identical rodents, the monogamous prairie vole and the promiscuous montane vole. The most notable neurological difference between the two species is the density and distribution of oxytocin receptors in their brains, the monogamous prairie vole enjoying greater densities of oxytocin receptors. Moreover, injections of oxytocin into appropriate brain areas lead the montane vole to preferences for a single partner over other partners, while injections of oxytocin blockers render the prairie vole less capable of monogamy. Other studies of voles find that oxytocin increases after sexual behavior, and that injections of oxytocin increase social contact and pro-social behavior, whereas blocking the activity of oxytocin prevents maternal behavior.

Studies of other species yield similar results. In primates, injections of oxytocin increase touching behavior and gaze focused on infants, and decrease threatening facial displays such as teeth-

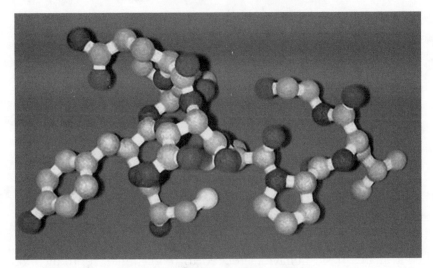

Your key to monogamy: oxytocin

baring yawns. Little domestic chicks, when separated from their mother, emit fewer separation distress calls after they have been given a dose of oxytocin. Oxytocin injections cause ewes to become attached to unfamiliar lambs.

Right now I suspect you're asking three questions. What about humans, that most complex of the pair-bonding species? Isn't oxytocin what Rush Limbaugh was addicted to? (No, that was OxyContin, an opioid painkiller; one wonders what his show would've been like had he grown addicted to oxytocin.) And where can I get this oxytocin and sprinkle it on my partner's morning Corn Flakes?

With respect to the first question, the literature on oxytocin in humans is beginning to reveal an equally compelling picture of the physiological underpinnings of love, devotion, and trust. In studies of lactating women, it has been found that oxytocin reduces activity of the hypothalamic-pituitary-adrenal axis, the physiological basis of stress. Prepartum mothers who show higher baseline levels of oxytocin later show increased attachment-related behavior with their new babies. Oxytocin is released in response to pleasurable massage and sex. Even chocolate triggers oxytocin release. It's not a coincidence that we give chocolate to loved ones on Valentine's Day, and not pickles, Pringles, or salsa—we're seeking to stimulate that feeling of trust and devotion.

So what about romantic love? To more directly document the relationship between oxytocin and romantic love, Gian Gonzaga and I undertook a Darwinian study of sexual desire and romantic love. Gian first did what a good descendant of Darwin does: He turned to Darwin's own observations. Darwin identified three kinds of love—"maternal love," "love," and "romantic love" (see table below)—which closely parallel our designations of caretaker love, romantic love, and desire, although Darwin uses the term "romantic love" to refer to what we now call sexual desire.

DARWIN'S DESCRIPTIONS OF THE VARIETIES OF LOVE

MATERNAL LOVE	TOUCH, GENTLE SMILE, TENDER EYES
LOVE	BEAMING EYES, SMILING CHEEKS (WHEN SEEING OLD FRIEND), TOUCH, GENTLE SMILE, PROTRUDING LIPS (IN CHIMPS), KISSING, NOSE RUBS
ROMANTIC LOVE	BREATHING HURRIED, FACES FLUSHED

Then Gian, envy of his graduate-student peers, sequestered himself in the depths of the library stacks and surveyed dozens of not-so-lurid studies of the nonverbal displays that accompany sexual intercourse versus friendly, affectionate contact in humans and nonhuman primates. What he identified were possible display behaviors that signal sexual desire and romantic love. Prior to sex, human and nonhuman primates tend to engage in a variety of lip- and mouth-related behaviors—they pucker lips, kiss, lick their lips, and stick their tongue out, stock in trade for rock and rollers like Mick Jagger.

In contrast, romantic love tends to be signaled with a warm, eye-glistening smile, a head tilt, and open-handed gestures. It is surprising that Darwin missed the open-handed gesture as a signal of love, for it is so readily explained by his principle of antithesis: We signal anger with clenched fists, tightened shoulders, flexed arms—the upper-body posture of the readiness to attack. Love, by implication, should be conveyed by the opposite—relaxed shoulders, head tilt, and open-handed gestures. It is no wonder that around the world

Facial cues of sexual desire.

greeting rituals between strangers employ open-handed gestures—signs of trust and cooperation. Our primate relatives, the chimps, resort to open-handed gestures to short-circuit aggressive tendencies, and to stimulate close proximity, grooming, and affiliation.

In our first study, we had young romantic partners come to the lab and talk about experiences of love and desire. These young partners, in love for an eternity—eighteen months—talked for a few minutes about when they fell in love. There were stories of meeting in a chemistry lab at 3:00 AM, of bumping into each other when skateboarding, of being charmed by the other's Facebook entries. And there in plain sight via intensive, frame-by-frame analysis, were four- to five-second bursts of the displays—flurries of lip licks, puckers and lip wipes, a smooth unfolding of smiles, head tilts and open palms. Our question was whether these brief behaviors, just a

The human display of romantic love; an open-handed display of chimps reconciling after a conflict.

few seconds long, would map onto distinct experiences of sexual desire and romantic love.

That indeed is what we found, and so much more. The brief displays of love increased as the partners, males and females alike, reported feeling more love at the end of the two-minute conversation. These microdisplays of love were unrelated to reports of desire. The brief displays of sexual desire, in contrast, correlated with the young lovers' reports of sexual desire, but not with love. Partners attributed more love, and not desire, to their partner when their beloved displayed more smiles, head tilts, and open-handed gestures; and they attributed greater desire when they saw their partner show those lip licks and lip puckers. In two-minute conversations, by carefully measuring half-second-long lip puckers and head-tilting smiles, we could pull apart these two great passions—romantic love and desire.

With further exploration, we uncovered other findings that may just change how you look at that partner across the dinner table from you. The couples who showed more intense nonverbal displays of love reported higher levels of trust and devotion and were more likely to have done something unusual for twenty-year-olds—to have talked about getting married. The couples who were swept away in desire were less likely to have talked about a future together (it gets in the way of desire) and reported less long-term commitment to one another. With this knowledge, I am ready for the stormy adolescences of my daughters. When their first dates come over or declare their romantic intentions, I am armed with the precise knowledge that I need. If I see a few too many lip licks and lip puckers as plans for the evening are discussed, it's a firm hand on the neck and a polite escorting out of our house.

We next turned to a query of our chemical quarry, oxytocin. Gian, Rebecca Turner, and I had women, who on average have seven times the rate of oxytocin in the bloodstream as men (oh, well), talk about an experience involving intense feelings of warmth for another person. As they recounted these experiences, blood was drawn and oxytocin was assayed some fifteen minutes later. From videotapes of these remembrances, we coded head-tilting smiles and open-handed gestures, as well as lip licks, puckers, and tongue

Each summer I teach Darwin and emotional expression at Kirk Cooper's *Sees the Day* summer camp. Here are four nine- and ten-year-olds' attempts at love, unprompted by any description. I'll leave it to your discerning eye to reach a conclusion about whether the girls or boys are doing a better job at it.

protrusions. Only the warm smiles, head tilts, and open-handed gestures increased with oxytocin release. The cues of sexual desire had nothing to do with the release of this neuropeptide of devotion and long-term commitment. The fulcrum on which marriage tips may be nothing more than these molecules of monogamy.

TRUST

In her cultural history *Dancing in the Streets*, Barbara Ehrenreich details humans' irrepressible tendency to dance, to move in rhythm toward collective joy and a love of one another. Paintings of dance are found on the earliest human pottery. Dance is part of many great myths, most notably the Maenads' celebration of Dionysus. It was a regular, ritualized occurrence of hunter-gatherer life. Dance may be the one uniformity, outside of eating, to collective gatherings—sporting events, political rallies, family reunions, religious meetings.

The early Christian church took an immediate dislike to communal dance—it generated subversive passions and could quickly sow the seeds of dissent and protest. Not surprisingly, the powers that be in the church (I suspect they had little rhythm) set in place extreme restrictions upon this human universal. But that proved to be, and will always be, a losing endeavor. The instinct to dance reemerged outside church walls in the form of carnivals, which persist to this day. Dance will emerge in any context, in church, at the game, at scholarly conferences, in strangers waiting in line for a bus, in two-year-olds bouncing to the beat of big bands at formal weddings. People need to sway their hips, shimmy their shoulders, and clap their hands together.

Our conceptual mistake, Ehrenreich observes, and it is a common one, is to assume that dance is sexual. Certainly our early, memorable experiences of dance—the intense slow-dance clutches to *Stairway to Heaven* of my eighth-grade youth—felt sexual. (Of course, anything in eighth grade is sexual—algebra, spelling bees, fire drills, corn dogs served at lunch.) But to generalize from these experiences to a broad statement about dance is misguided.

Instead, dance creates a love for fellow group members; it coordinates evolved patterns of touch, chant, smiling, laughing, and head shakes to spread collective joy in the sweat and delirium of collective movement. Dance is the most reliable and quickest route to a mysterious feeling that has gone by many names over the generations: sympathy, agape, ecstasy, *jen*; here I'll call it trust. To dance is to trust.

If neuroeconomist Paul Zak could study the neural correlates of that particular kind of love—of fellow group members—that rises after a great bout of dancing, he would likely find oxytocin levels shooting through the roof. Zak proposes that oxytocin is the biological underpinning of trust—a thesis he has supported in his groundbreaking work with the trust game. In the trust game, one participant, known as the "investor," makes contributions to another individual, known as the "trustee." The value of the money given to the trustee then triples, and the trustee then gives some amount back to the investor—as much or as little as he or she desires. As in so many realms of life, cooperation amplifies the potential gains to be had by all, but it requires a leap of faith, a core conviction, a sense of trust, that the trustee will give back some of the funds generously given.

In studies Zak has conducted in Germany and Switzerland (where it is not illegal to study oxytocin experimentally) Zak has given a blast of oxytocin, or a neutral solution, to the investor via a nasal spray. Our "investor," grooving on oxytocin, was more than twice as likely to give away maximum amounts of money to the stranger than the "investor" given a neutral solution in the control condition (see figure below).

My former student Belinda Campos calls this cocktail of love toward non-kin, enhanced by oxytocin and founded in the sense of trust, the love of humanity. Her research shows that this feeling, and not other kinds of love, amplifies the conviction in the goodness of other humans. It is accompanied by the urge to give, to trust, and to sacrifice. In one study we examined college students' transitions to their new community—their residence hall—during their first year of college. Students who reported feeling a great deal of the love of humanity prior to coming to college more quickly trusted their new hallmates, and folded more quickly into dense webs of friendships. It is the feeling that led Gandhi to say that "all men are brothers" and Jesus to say "Whoever love all brothers has obeyed the whole law" (Romans 13:8–10). It is the love of humanity that weaves together Walt Whitman's declarations in *Song of Myself*.

And empirical studies are finding that the health of communities depends on trust and the love of humanity. Robert Sampson, at Harvard University, has found that in resource-deprived, dangerous neighborhoods, children fare better when they feel a sense of love of humanity from their neighbors. In these neighborhoods, adults who make warm eye contact with neighborhood children, who provide that comforting pat on the back, who speak with encouraging words and in uplifting tones, create a sense of trust and strength in the young non-kin in their midst. In other research on divorce and the fractured family, children prove to be much more resilient in the wake of their parents' divorce when they feel a sense of connection to and devotion for other nearby adults— neighbors, teachers, coaches, pastors.

In the small groups in which we evolved, there were few walls that separated kin from non-kin. All were likely engaged in the sharing of caretaking behavior, cooperation in gathering resources, defense against predation. Our success at these tasks hinged critically upon a sense of trust in others, on the emergence of a love of

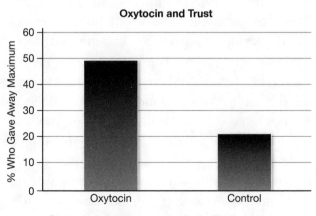

Oxytocin increases generosity in the trust game.

humanity. Evolution responded with a deeply rooted set of behaviors related to love and trust—feelings of devotion, the urge to sacrifice, a sense of the beauty and goodness of others, affectionate touch, oxytocin, activation in the reward circuitry of the brain, the shutting down of the threat circuitry of the brain (the amygdala), mutual smiles and head tilts, open-handed gestures and posture, a soft, affectionate tone in the voice. These in their earliest forms were most evident in the early attachment dynamics of parent and child and in the quiet, isolated moments of intimacy between reproductive partners. These patterns of behavior were readily spread to non-kin, in rituals like dance and feast, serving as the basis for friendships. They spread informally, through the contagious power of these emotions. Passing a young, bundled-up baby from mother to friend, in a common exchange of caretaking, might bring about shared coos, smiles, and cradling of the child, and so much more—a sense of community.

BACK TO THE BIRDS AND THE BEES

If I could have taken another shot at helping my daughters understand the realm of love in the wake of the elephant seal disaster, I might have tried to walk them through the figure below. This figure portrays what social science has found about the varieties of love across a human's life. Perhaps I would have started with the story of the Wild Boy of Aveyron, and the corresponding science showing that in humans, loving relationships (of any kind) lead to less depression and anxiety, greater happiness, more ruddy health, a more robust nervous system, and greater resistance to disease (not to mention it just feels good). I would have told Natalie and Serafina that as they age, and as the end nears, psychologist Laura Carstensen has found time and time again, loving relations get more important, and love all the sweeter. So why not start now?

I would have told them that the love between parent and child (the dark solid line) fluctuates; it dips necessarily during adolescence when they themselves (or their children, some twenty years from now) will be throwing themselves into romantic relations of their own. They shouldn't be alarmed when this happens to them (although I'm certain I will be more than alarmed); the love of caretakers and those we take care of returns and branches into the delightful love of grandparents for grandchildren. The circle expands.

I would have told them of the delights of that most intense of loves, passionate love (the dark dotted line), and of its head-spinning, heart-pounding delirium, but that we mustn't be tricked for too long by its celestial charms. When passionate desire dips postchildbirth, in particular between year one and four into life with young children, researchers find, romantic relations become vulnerable. As it declines through the life course, as much as we (or the multibillion-dollar beauty industry) might think otherwise, other forms of love become so much sweeter.

I might have cautioned that after the golden period of romantic love (the gray solid line), which they are too soon to head into, romantic love dips during the early years of raising children, overshadowed by demands such as spit-up, phone tag over playdates, and temper tantrums. I would remind them of the love that reemerges in the empty nest. I would ask them to read Stephanie Coontz's *History of Marriage*, where she suggests the one mistake we make today in marriage is to put too much of a burden on romantic love; that we need more diverse kinds of love. I would

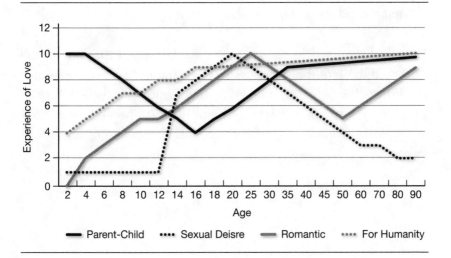

refer to the new science of relationships, which suggests that romantic love does not live on passion alone. It requires many other positive emotions—laughter, play, a sense of wonder, kindness, forgiveness—to arrive at that magic ratio of five positive feelings for every toxic negative one that enables marriages, in John Gottman's wisdom, to endure.

And I would have tried to describe the love of humanity, agape, really a love of all sentient beings (the gray dotted line). This feeling is the central discovery—the heart, so to speak, of ethical systems ranging from Tibetan Buddhism to major strands of Christianity. It is a love that generates trust, generosity, and stable communities. It is the ether in the air of peaceful playgrounds, Sunday strolls in the park, quiet reverence in museums and churches. It may be the clue toward beating things like global warming. It is a kelson of creation in Walt Whitman's *Leaves of Grass*:

> *Swiftly arose and spread around me the peace and joy and*
> > *knowledge that pass all the art and argument of*
> > *the earth;*
> *And I know that the hand of God is the elderhand of my own,*
> *And I know that the spirit of God is the eldest brother*
> > *of my own,*
> *And that all the men ever born are also my brothers . . .*
> > *and the women my sisters and lovers,*
> *And that a kelson of the creation is love.*

I would have tried to convey that their lives, and their children's lives, and those of their friends, and the character of the communities they will inhabit, are shaped by their search for these four passions. I would have wished them well in hoping that life's arrangements would allow for the fullest expression of these four loves.

11

Compassion

ONE DAY WHILE FIGHTING in the Spanish civil war, George Orwell encountered an enemy Fascist face-to-face. The soldier came running by, panting, half dressed, stumbling, holding up his pants with clenched hand. Orwell refused to shoot. Later he reflected: "I did not shoot partly because of that detail about the trousers. I had come here to shoot at 'fascists'; but a man who is holding up his trousers isn't a 'Fascist,' he is visibly a fellow-creature, similar to yourself, and you don't feel like shooting at him." The sight of the Fascist's bare chest, his skin, his disheveled condition, had short-circuited Orwell's instinct to kill.

In *Humanity*, historian Jonathan Glover documents many such "sympathy breakthroughs" in the wars of the twentieth century—in Stalin's purges, the My Lai massacre, the killing fields in Cambodia, the genocide in Rwanda. These are moments when soldiers break free from the dutiful honoring of the military code, from strict orders to shoot on sight, and are overwhelmed by the humanity of the humans they are killing. Most often it is when encountering children and women—for example, the toddlers and pregnant women beheaded and disemboweled in the My Lai massacre. Most often the sympathy breakthroughs are triggered by eye-to-eye contact, the sight of the enemy's pupils, the pores on his skin, oblique movements in his eyebrows.

No sympathy breakthrough was more dramatic than that of Miklós Nyiszli, a medical doctor at a Nazi concentration camp. One day as a gas chamber was being cleared of bodies, a young girl of sixteen was found alive at the bottom of the rigor mortis pile of thin-limbed, stiffening corpses. The attending staff reflexively offered the young girl an old coat, warm broth, tea, and reassuring touch to her shoulders and back. Nyiszli tried to persuade the concentration camp's commandant to save her. One proposal was to hide her amid German women working at the camp. The commandant toyed with this possibility momentarily but in the end had her killed by his method of choice—the young girl was shot in the back of the neck.

Human history, Glover contends, can be thought of as a contest between cruelty and compassion, tellingly revealed in wartime sympathy breakthroughs, when the force of compassion overwhelms the edicts of war. You could make the same case about human nature. Fight/flight tendencies of self-preservation are continually at odds with tendencies to care in the electrochemical flow of our nervous systems. The content of the mind shifts between the press of self-interest and the push of compassion. The ebb and flow of marriages, families, friends, and workplaces track the dynamic tension between these two great forces—raw self-interest and a devotion to the welfare of the other. The study of emotion is experiencing its own "sympathy breakthrough" thanks to recent studies of compassion, which are revealing this caretaking emotion to be built into our nervous systems. The study of this emotion holds new clues about the health of marriages, families, and communities.

THE COMPASSION CONSPIRACY

As Charles Darwin developed his first account of the evolution of humans in the *Descent of Man*, he argued for "the greater strength of the social or maternal instincts than that of any other instinct or motive." His reasoning was disarmingly intuitive: In those collectives of our hominid predecessors, communities of more sympathetic individuals were more successful in raising healthier offspring to the age of viability and reproduction—the surest route to getting genes to the next generation, the *sine qua non* of evolution.

Darwin's elevation of sympathy as the strongest of our instincts, and as the foundation of ethical systems, has not attracted many adherents in the annals of Western thought. More typically, sympathy and compassion have been treated with dismissive skepticism or downright derision. Thomas Huxley argued that evolution did not produce a biologically based capacity to care; instead, kindness, cooperation, and compassion are cultural creations, constructed within religious commandments and rituals, in norms governing public exchange, codified in social organizations, as desperate attempts to rein in, to countervail man's base tendencies. The regularity of parents abandoning and abusing children, infanticide, torture, and genocide lend compelling, if not overwhelming, credence to Huxley's counterpoint. Scientists searching for an evolved, biological basis of compassion, by implication, would be grasping at the air, tilting their labs at windmills.

Other influential thinkers in the Western canon, reveals philosopher Martha Nussbaum in her brilliant history of the study of emotion in *Upheavals of Thought*, have gone further. The trend in Western thought has been to argue that compassion is an unreliable guide to ethical behavior (see quotations below). Compassion is "blind," too subjective to be a universal guide to the conscience and ethical conduct. It is imbued with the individual's idiosyncratic concerns (what's unwarranted suffering in my eyes is justified in yours). Compassion is "weak"; it enfeebles the individual in the hard work of meting out justice.

A feeling of sympathy is beautiful and amiable; for it shows a charitable interest in the lot of other men. . . . But this good-natured passion is nevertheless weak and always blind.
IMMANUEL KANT, *Observations on the Feeling of the Beautiful and Sublime*

If any civilization is to survive, it is the morality of altruism that men have to reject.
AYN RAND, *"Faith and Force: The Destroyers of the Modern World"*

A transvaluation of values, under the new pressure and hammer of which a conscience should be steeled and a heart transformed into brass, as to bear the weight of such a responsibility.
FRIEDRICH NIETZSCHE, *Beyond Good and Evil, section 203*

Hence a prince who wants to keep his authority must learn how not to be good, and use that knowledge, or refrain from using it, as necessity requires.
MACHIAVELLI

These old notions have blinded the scientific study of compassion. New empirical studies, though, have mushroomed, and yet again give the nod to Darwin. Compassion is a biologically based emotion rooted deep in the mammalian brain, and shaped by perhaps the most potent of selection pressures humans evolved to adapt to—the need to care for the vulnerable. Compassion is anything but blind; it is finely attuned to vulnerability. It is anything but weak; it fosters courageous, altruistic action often at significant cost to the self. These discoveries would be founded upon the study of a region of the nervous system that has remained mysterious to scientific understanding until recently.

LOST VAGUS

In calling sympathy the strongest of instincts, Darwin was touching a nerve in the veins of canonical Western thought. Little did he know, Darwin was also touching another nerve, literally a bundle of nerves, known as the vagus nerve, which resides in the chest and, when activated, produces a feeling of spreading, liquid warmth in the chest and a lump in the throat. The vagus nerve (see figure below) originates in the top of the spinal cord and then winds its way through the body (*vagus* is Latin for wandering), connecting up to facial muscle tissue, muscles that are involved in vocalization, the heart, the lungs, the kidneys and liver, and the digestive organs. In a series of controversial papers, physiological psychologist Steve Porges has made the case that the vagus nerve is the nerve of compassion, the body's caretaking organ.

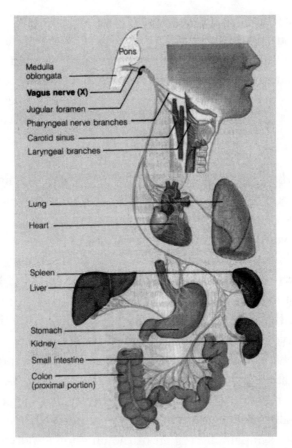

The vagus nerve

How so? First of all, Porges notes that the vagus nerve innervates the muscle groups of communicative systems involved in caretaking—the facial musculature and the vocal apparatus. In our research, for example, we have found that people systematically sigh—little quarter-second, breathy expressions of concern and understanding—when listening to another person describe an experience of suffering. The sigh is a primordial exhalation, calming the sigher's fight/flight physiology, and a trigger of comfort and trust, our study found, in the speaker. When we sigh in soothing fashion, or reassure others in distress with our concerned gaze or oblique eyebrows, the vagus nerve is doing its work, stimulating the muscles of the throat, mouth, face, and tongue to emit soothing displays of concern and reassurance.

Second, the vagus nerve is the primary brake on our heart rate.

Without activation of the vagus nerve, your heart would fire on average at about 115 beats per minute, instead of the more typical 72 beats per minute. The vagus nerve helps slow the heart rate down. When we are angry or fearful, our heart races, literally jumping five to ten beats per minute, distributing blood to various muscle groups, preparing the body for fight or flight. The vagus nerve does the opposite, reducing our heart rate to a more peaceful pace, enhancing the likelihood of gentle contact in close proximity with others.

Third, the vagus nerve is directly connected to rich networks of oxytocin receptors, those neuropeptides intimately involved in the experience of trust and love. As the vagus nerve fires, stimulating affiliative vocalizations and calmer cardiovascular physiology, presumably it triggers the release of oxytocin, sending signals of warmth, trust, and devotion throughout the brain and body and, ultimately, to other people.

Finally, the vagus nerve is unique to mammals. Reptilian autonomic nervous systems share the oldest portion of the vagus nerve with us, what is known as the dorsal vagal complex, responsible for immobilization behavior: for example, the shock response when physically traumatized; more speculatively, shame-related behavior when socially humiliated. Reptiles' autonomic nervous systems also include the sympathetic region of the autonomic nervous system involved in fight/flight behavior. But as caretaking began to define a new class of species—mammals—a region of the nervous system, the vagus nerve, emerged evolutionarily to help support this new category of behavior.

Historians of science have rated Charles Darwin as off-the-charts in terms of kindness and warmth relative to other groundbreaking scientists (he was the only passenger on the *Beagle* about whom not a negative word was said, and was friend to captain and ship hands alike). As Darwin wrote in his Down House, amid the noisy, loving spectacle of his ten children, he most certainly felt those sensations of expansive warmth in the chest associated with the vagus nerve. The humming of the vagus nerve may have led Darwin to his often neglected thesis that sympathy and the maternal instincts are the centerpiece of human social evolution, that they bring the good in others to completion and are a foundation of high *jen* ratios. Some hundred and thirty years later a new science has yielded similar insights.

NERVES OF COMPASSION

Steve Porges's wild-eyed claims about the vagus nerve would have inspired William James. James was the progenitor of the notion that our emotions originate in patterned responses in the autonomic nervous system, which lies below the brain stem and coordinates basic tasks like the distribution of blood, digestion, sexual response, and breathing. What could be more compelling proof that our emotions are embodied in peripheral physiological response, "reverberations of the viscera," in James's Victorian language, than the notion that that loftiest of human emotions—compassion—has its own bundle of nerves located deep within the chest?

Walter Cannon, a student of William James's, was not so convinced by his advisor's provocative armchair musings. The responses of the autonomic nervous system, Cannon countered, do not carry enough specific meaning to account for the many distinctions people make in their emotional experience. Patterned changes in heart rate, breathing, goose bumps, pupil dilation, cotton mouth, and sweaty palms could never give rise to nuances in experiences of gratitude, reverence, compassion, pity, love, devotion, desire, and pride.

On top of that, Cannon continued, the autonomic responses of emotion are simply too slow to account for the rapidity with which we experience emotion or move from one emotion to another. The autonomic nervous system typically produces measurable responses within fifteen to thirty seconds after the emotion-eliciting event. Clearly our emotional experience arises more rapidly. The blush, for example, peaks at about fifteen seconds after the embarrassing event; our experience of embarrassment, in contrast, arises immediately upon the recognition of the mistake we have made. In Cannon's eyes, the autonomic nervous system is too slow-moving a system to account for the meteoric emergence and nimble shifting of our emotional experiences.

Finally, and perhaps most persuasively, we are relatively insensitive to the changes in the autonomic nervous system—heart rate increases, sweaty palms, vasoconstriction in the veins of your arms or legs, blushing, or activity in your intestines. Cannon noted, for example, that people actually feel little when their intestines are cut or burned. Slightly less dramatic empirical studies have found that when people are asked to guess whether their heart rate has increased or decreased, they most typically fare little better than chance. Even if the autonomic nervous system generated emotion-specific responses, it is not clear that we would perceive these bodily changes with our conscious minds. It is even less clear to assume that these dimly perceived bodily sensations would weave their way into our emotional experience. It would be foolhardy, by implication, to seek to locate all the nuances of compassion—the sense of undeserved harm, the feeling of concern and common humanity, the urge to help—in something so diffusely distributed in the peripheral nervous system as the vagus nerve.

Undaunted, my student Chris Oveis has risked his career on the very hypothesis that the vagus nerve is a bundle of caretaking nerves. He did so by starting in an obvious place—suffering. Humans are wired to respond to harm from the first moments of life. One-day-old infants cry in response to another infant's cries of distress but not their own. Many two-year-old children, upon seeing another cry, will engage in the purest forms of comfort, offering their toys and gestures of visible concern to the person suffering. Pictures of sad faces presented so fast participants don't even know what they've seen trigger activation in the amygdala.

So we asked first whether the exposure to harm would trigger activation in the vagus nerve, and whether an emotion that revolves around the inclination to distance oneself from weak others—pride—would not. In the compassion condition, participants viewed images of malnourished children, suffering during wartime, and infants in distress, images that fit the Aristotelian notion of the purest elicitors of compassion: another's suffering that is extreme and undeserved. Participants in the pride condition viewed slides that would arouse the pride of our UC Berkeley undergraduates—images of landmarks on the campus, pictures of Cal sporting events,

A slide used to induce compassion.

A slide used to induce pride.

and, perhaps the most inspiring of all, a picture of the Cal mascot, Oski the bear.

As participants viewed 2.5 minutes of slides, we measured activity of the vagus nerve with electrodes attached to the chest and a band placed around the abdomen to measure breathing. These measures yield an index called respiratory sinus arrhythmia (RSA), which has been developed in the past fifteen years to capture activation in the vagus nerve. RSA works as follows. When we inhale, the vagus nerve is inhibited, and heart rate speeds up. When we exhale, the vagus nerve is activated, and heart rate slows down (which is why so many breathing practices prioritize exhaling and are soothing to the soul—and a source, perhaps, of compassion). The vagus nerve controls how breathing influences fluctuations in heart rate. We measure the strength of the vagus nerve response, therefore, by capturing how heart rate variability is linked to cyclical changes in respiration.

The first finding of importance from Chris's study was that brief exposure to images of harm triggered activation of the vagus nerve more so than the images that made participants proud. Perhaps more convincingly, participants' experiences of compassion and pride were, as James would have hypothesized, quite sensitive to fluctuations in the activity of the vagus nerve. Participants' reports of their feelings of compassion increased as their vagus nerve activity increased; participants' self-reports of pride decreased as their vagus nerve activity increased. With increasing vagus nerve response, participants' orientation shifted toward one of care rather than attention to what is strong about the self.

Then our participants, feeling surges of either compassion or pride, indicated how similar they themselves were to twenty other groups. They rated their common humanity with Democrats, Republicans, saints, small children, convicted felons, terrorists, the homeless, the elderly, farmers, and, God forbid, Stanford students. Why this odd task? To ascertain whether compassion shifts people's sense of similarity to others—a potent enabler of altruistic action. Philosopher Peter Singer has argued that this sense of similarity, or circle of care, is a core ethical principle that emerged as part of the evolution of the ethical mind. In Singer's words, evolution has

> bequeath(ed) humans with a sense of empathy—an ability to treat other people's interests as comparable to one's own. Unfortunately, by default we apply it only to a very narrow circle of friends and family. People outside that circle were treated as subhuman and can be exploited with impunity. But over history the circle has expanded . . . from village to the clan to the tribe to the nation to other races to other sexes . . . and other species.

This expanding circle of care gives rise to a belief in equality, to the extension of individual rights to others. It is the target of many meditation exercises, which discipline the mind to come to treat all sentient beings with loving kindness. It is advocated by spiritual leaders, from the Buddha to Jesus. It is at the heart of *jen*. And it is a deep intuition that is intertwined with activity of the vagus nerve

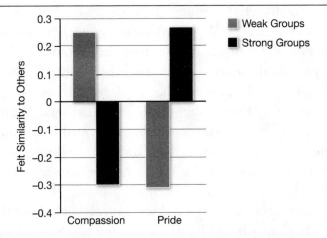

Compassion makes people feel similar to weak groups; pride makes people feel similar to strong groups.

in the depths of the human chest. Our participants made to feel compassion by viewing images of harm reported a broader circle of care—they reported a greater sense of similarity to the 20 groups—than people feeling pride. This feeling of similarity to others increased as individuals' vagus nerve fired more intensely. And when we looked more closely at whom people feeling compassion and pride felt most similar to (see figure above), we found that pride made people feel more similar to the strong, resource-rich groups in the set of twenty they rated—Berkeley and Stanford undergraduates, lawyers, and the like (the dark bar to the far right). Compassion, on the other hand, made people feel more similar to the vulnerable groups—the homeless, the ill, the elderly (the gray bar to the far left). Compassion is anything but blind or biased by subjective concerns; it is exquisitely attuned to those in need.

ALTRUISM'S HOLY GRAIL

There are theoretical cottage industries devoted to attributing seemingly altruistic action to selfish motivations. Take Paul Rusesabagina's remarkable heroism during the genocide of Rwanda, so powerfully depicted in Philip Gourevitch's *We Wish to Inform You That Tomorrow We Will be Killed with Our Families* and the film *Hotel Rwanda*. Rusesabagina risked his own life, and that of his wife and children, to save hundreds of Tutsis (he is a Hutu) from the genocidal Hutu militia, the Interahamwe, sheltering them at the hotel Milles Collines, which he managed. Within the social sciences, these courageous actions are readily attributed to selfish genes, to the desire to save kin, or to self-interest, pure and simple. Freudian-leaning theorists have also weighed in—altruistic action is a defense mechanism by which we ward off deeper, unflattering, anxiety-producing revelations about the self ("If I give to charity then I'll think less about how much I hate my father!"). The more parsimonious account—that Paul Rusesabagina, and we on our best days, act altruistically because we are wired to care for others—plays second fiddle to selfish accounts of altruism in this age-old debate about the origins of goodness.

In an essay on the sublime and the beautiful, Immanuel Kant zeros in on the possibility that compassion renders people weak and passive in the face of injustice. Digressing somewhat, Kant observed:

> For it is not possible that our heart should swell from fondness for every man's interest and should swim in sadness at every stranger's need; else the virtuous man, incessantly dissolving like Heraclitus in compassionate tears, nevertheless with all this goodheartedness would become nothing but a tender-hearted idler.

Compassion turns people into passive, timid, melancholic sorts, "tender-hearted idlers" like the philosopher Heraclitus, known for his thesis that human nature is always in flux. We can thank Daniel Batson and Nancy Eisenberg for taking on this deeply entrenched claim, and gathering empirical data that show that compassion is the holy grail of altruism researchers—a pure, other-oriented state that motivates altruistic behavior like that which Paul Rusesabagina so courageously displayed during the genocide in Rwanda.

To set the stage for his empirical studies, Batson argues that any noble or altruistic act can have multiple motives. Seemingly altruistic actions—donations to charity, staying late to help a colleague, climbing a tree to rescue a child's kitten, helping an elderly woman cross an icy street—are often driven by selfish motives. One such selfish motivation is to reduce the distress we ourselves feel at the sight of another person suffering (it is still quite remarkable that we suffer at the sight of another suffering). A second is the allure of social praise—we help those in need to win those gold stars in the classroom, the Boy Scout badges, public service awards, approving head nods of parents, and to burnish our reputations in the eyes of our peers.

Batson also maintains, in theorizing that would have warmed Darwin's heart or more precisely his vagus nerve, that there is an other-oriented state that can be the wellspring of altruistic behavior: compassion. The question is how to document that this selfless state of compassion produces altruism. Batson's solution to this challenge is to put people in experiments where they are confronted with someone in need, and their experience of compassion and selfish motives—for example, to slip out of helping with little notice—clash. If one observes altruistic action in this clash of selfish motives and compassion, we can infer that compassion won the day and motivated the altruistic action. It's a bit like testing a new lover by allowing him or her possibilities of intimate affection from others. If in this clash of competing affections he or she returns, faithful, doughy-eyed and devoted, one has learned about commitment.

In a first study, Batson had participants watch another participant (actually a confederate) complete several trials of a memory task. After each mistake, this individual received—of course—a wince-producing, shoulder-jolting shock. In one condition, the easy-escape condition, the participant was only required to watch the confederate receive two of the ten shocks. At that moment, the participant was free to leave. Here the participant should be guided by the selfish inclination to reduce personal distress in witnessing the other person suffer; all the participant had to do was leave. In the difficult-to-escape condition, the participant had to watch the other person take all ten shocks.

After the first two trials, the individual receiving the shocks began to look a little pale. He mumbled for a glass of water. He mentioned feelings of discomfort and recounted a traumatic shock experience from childhood. While the experimenter figured out how to respond to this complex turn of events, the participant reported on how distressed and compassionate he or she was feeling at that moment. Then the experimenter hit upon an idea: Would the participant take some of the shocks on behalf of the other individual, clearly dreading the prospects of more shocks? The critical test for our present interests was in the condition in which participants were feeling compassion but were allowed to leave. Which branch of their nervous system prevailed—the selfish or the compassionate? The compassionate. These participants, feeling the swell of compassion in their chest but hearing the voice of pure self-interest—they could just pick up and leave—volunteered to take several more shocks on behalf of the other participant.

One worry you might have is that participants who took more shocks did so simply to impress the experimenter. Fair enough, a reasonable critique. Does compassion drive altruistic action even in purely anonymous settings? Does the absence of the opportunity to gain social rewards—the esteem of others—dampen our nobler inclinations? This age-old question motivated Batson's next study. In this study, female participants conversed with another participant (a confederate) through an exchange of notes while seated in separate cubicles. Some participants were told to be as objective as possible when reading the notes, to concentrate on the facts at hand. Other participants were asked to imagine as vividly as possible how the communicator—the other person—felt; they were led to feel compassion.

The first note the participant read was from a student named Janet Arnold, who confessed to feeling out of place at her new home at the University of Kansas. She hailed from the rolling hills of nearby Ohio and was having a bit of difficulty adjusting to the exotic locale of Lawrence, Kansas. In the second exchange, Janet expressed a strong need for a friend. She asked the participant, point-blank, if she'd like to hang out together. Upon reading this second note, the participant was told that Janet had finished and left the study and was then asked to indicate how much time she would be willing to spend with Janet. Her response would be read by Janet and the experimenter or it would remain anonymous. The individual who volunteered to spend the most time with Janet? The person who was feeling compassion and in the anonymous condition.

Stronger evidence still would link selfless, altruistic action to activation in the vagus nerve. Nancy Eisenberg has gathered just this kind of data. In one illustrative study, young children (second-graders and fifth-graders) and college students watched a videotape of a young mother and her children who had recently been injured in a violent accident. Her children were forced to miss school while they recuperated from their injuries in the hospital. After watching the videotape, the children were given the opportunity to take homework to the recovering children during their recess (thus sacrificing precious playground time). Those children who reported feeling compassion and who showed heart rate deceleration—a sign of vagus nerve activity—as well as oblique, concerned eyebrows while watching the video (see figure below) were much more likely to help out the kids in the hospital. In contrast, those children who winced, who reported distress, and who showed heart rate acceleration—that is, those children who reacted with their own personal distress—were less likely to help. These findings make a clarifying point: It is an active concern for others, and not a simple mirroring

The oblique eyebrows and lip press of sympathy, which predict that altruistic behavior will occur.

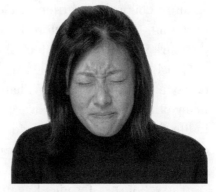

The painful wince that predicts the turning away from those who are suffering.

of others' suffering, that is the fount of compassion, and that leads to altruistic ends.

These scientific studies countervail the influential claims of the Kants, Nietzsches, and Rands about the nature of human goodness. Compassion is not a blind emotion that catapults people pell-mell toward the next warm body that walks by. Instead, compassion is exquisitely attuned to harm and vulnerability in others. Compassion does not render people tearful idlers, moral weaklings, or passive onlookers but individuals who will take on the pain of others, even when given the chance to skip out on such difficult action or in anonymous conditions. The kindness, sacrifice, and *jen* that make up healthy communities are rooted in a bundle of nerves that has been producing caretaking behavior for over 100 million years of mammalian evolution. And the lives of individuals with highly active vagus nerves add yet another chapter to the story of how we are wired to be good.

VAGAL SUPERSTARS

Our tendencies to experience specific emotions, fleeting and evanescent as they are, define who we are. Emotions shape our deepest beliefs and core values, our relationships, the careers we choose, our methods for handling conflict, the art we like, the foods that please us, the very trajectory of our lives and those of our spouses, children, and friends. Descartes did not quite get it right in stating, "I think, therefore I am"; he would have been more on the mark if he had said, "I feel, therefore I am."

Consider what has been learned about shyness—a temperamental style characteristic of William James, Virginia Woolf, and so many others who have uncovered the mysteries of emotion. Early in life shy individuals show evidence of a hyperactive fear system, or HPA axis, which shapes their patterns of relationships and life choices. We know this thanks to the longitudinal studies of Harvard psychologist Jerome Kagan. Kagan has identified very shy infants at four months of age according to their fearful, distressed reactions to novel toys. Fast-forward seven years to Kagan's observations of these children in social groups: Shy children identified at age four months are most likely to be those two or three children in grammar-school classes who hover at the edges of the playground, observing and analyzing rather than engaging in the pyrotechnic face-to-face dynamics of that age (my bet is that a disproportionate number of writers fit this profile). Shy children have stronger stress reactions (elevated heart rate, pupil dilation, cortisol response) when hearing fiction being read or when engaging in complex cognitive tasks. And these same individuals, at age twentyone, when in an fMRI scanner and presented with slides of faces they had not seen, show stronger activation in the amygdala. When Avshalom Caspi studied the adult lives of shy individuals, he found, fitting with the analysis here, that shy individuals took almost 2 additional years, compared to more outgoing types, to enter into marriage, and they also took longer to settle upon a stable job. That fearful 4 month old, startled and distressed at the presence of a new toy, fight or flight physiology throbbing in the veins and throughout the body, is likely to lead a life of restraint, inhibition, and hesitation in the face of intimacy.

If the vagus nerve is a caretaking organ, then one would expect individuals with elevated vagus nerve activity to enjoy rich networks of social connection, to show highly responsive caretaking behavior, and for compassion to be at the center of their emotional lives. New studies are finding this to be the case. In one study, Chris Oveis and I brought Berkeley undergraduates to our lab in October and had their vagus nerve activity measured (deriving a measure known as vagal tone) while they sat quietly and comfortably in a resting state. Our interest was in tracking the lives of people with elevated vagus nerve activity in a resting state—vagal superstars. When they returned to the lab seven months later, we found that our vagal superstars, compared to those individuals with low baseline vagal tone, reported elevated levels of the trait extraversion, which is defined by high levels of social energy, friendships, and social contacts, and agreeableness, which is defined by great warmth, kindness, and a love of others. People with elevated baseline vagal tone also reported more optimism, general positive mood, and better physical health seven months later. And when presented with images of harm and beauty, they reported greater compassion and awe—their minds were more active in the aesthetic realm.

Perhaps most dramatically, we found that the vagal superstars showed an increased propensity for transformative experiences of the sacred. Approximately three months after we had assessed baseline vagus nerve activity, we e-mailed our participants, asking them the following question: "While going through college, people sometimes have experiences that have an important impact on their sense of meaning and purpose, or how they see themselves or the world. Since coming in for your initial lab visit for this project, can you please describe any experiences of this kind you have had?" Sixty-five percent of participants reported such a transformative experience during the three-month period between the initial lab visit and the e-mail query. There were accounts of nature, of going to a political rally, of hearing an inspiring person speak about global warming or free markets, of relatives and friends passing away and the contemplation of death, of being engaged in spiritual practice. This age is a fertile time of expansion and transformation. Here are a couple of examples:

> "I went to winter camp with my church. We stayed in the mountains for four days. . . . There was a guest speaker there who gave a very powerful message on the last night. It made me feel like God had a plan for me."

> "After my father's passing, I pondered what is the purpose of life. It changed me in that I'm closer to my family and I'm more responsible than before."

When we coded these transformation narratives, the central theme that emerged was a shift toward increased connection with others, an inclination to sacrifice, to be altruistic. And, yes, our high vagal tone individuals were more likely to report this kind of transformative experience.

Elevated vagus nerve activity, then, orients the individual to a life of greater warmth and social connection. Nancy Eisenberg has found that seven- and eight-year-olds with a higher resting vagal tone are more helpful in class, more sympathetic to those in need, more pro-social toward their friends, and experience more positive emotions. College students with higher resting vagal tone are better able to cope with the stresses of college—exam periods, career choices, the vicissitudes of romantic life. Following the loss of a married partner, people with high resting vagal tone recovered more quickly from the depressive symptoms that often accompany bereavement. And on the other end of the continuum, people experiencing severe depression, and its accompanying impoverishment of social connection, have been shown to have low resting vagal tone.

If William James had been a psychophysiologist with a high-tech lab, and had been able to study the vagus nerve, I suspect he would have brought Walt Whitman in—an inspiration to James and a source of his writings on the optimistic and embracing spirit. James observed that Whitman was known by all to be uniformly kind, generous, and upbeat. Had James recruited Whitman as a participant, and hooked him up to electrodes near the heart and to the respiration band around his girth to derive an assessment of Whitman's baseline vagal tone, I bet he would have found his vagal tone to be stratospherically high, and to soar at each thought of the beauty of our species or the wonders of leaves of grass.

THE SPREAD OF SELFLESS GENES

The great shift in early hominid social organization had to do with the arrival of hypervulnerable, big-brained offspring. The success of getting genes to the next generation hinged in unprecedented ways on getting dependent offspring to the age of viability and reproduction—a hair-raisingly long thirteen or fourteen years. Our vulnerable offspring shifted the reproductive dynamics of females and males toward a pattern of serial monogamy. Our vulnerable offspring's need for care got fathers into the act—hominid fathers provide more care for offspring than almost all other primates. The vulnerability of our offspring, Sarah Blaffer Hrdy writes in *Mother Nature*, exceeded the capacities of any single parent, and thus necessitated cooperative systems of child rearing built upon trades and exchanges between kith and kin. It was take care or die for our early primate predecessors.

The profound vulnerability of our big-brained offspring wired into us an instinct to care. It created in us a biologically based capacity for sympathy. It produced a vagus nerve, loaded up with oxytocin receptors, the provenance of feelings of devotion, sacrifice, and trust. It yielded a rich set of signals—empathic sighs, oblique eyebrows, and soothing touch, which trigger vagus nerve response and oxytocin and opioid release in the recipient, giving rise to oceanic feelings of connection. It produced specific cells underneath the surface of the skin that fire in response to the slow, soothing touch of compassion. The selection pressure to take care produced the indescribably beautiful qualities of the offspring themselves, designed, as many have argued, to reset the parents' nervous systems toward more caretaking settings. When parents look at pictures of their new babies, the orbitofrontal cortex lights up, as does a region called the periaqueductal gray, a bundle of neurons known to coordinate the patterned actions of grooming in primates. So great is the evocative power of the baby that baby-faced cues in adults—big forehead, big eyes, small chin—trigger trust and liking in other adults, and short-circuit the tendency to punish (if you're on trial, you're well served by increasing the size of your forehead and eyes).

But evolution did not stop there. So critical was caretaking to the survival of our species that it was selected for in other ways, guaranteeing that the capacity to be kind would be woven into the genetic fabric of this new hominid. A first is through sexual selection, the processes, initially described by Darwin, according to which certain individuals prevail in competitions with their own sex to gain access to mates, thereby gaining reproductive opportunities and increasing the likelihood of passing on their genes to the next generation. What sorts of people prevail in the meat markets, singles bars, speed dating and online dating services, and more run-of-the-mill matchmaking of modern life? Full-lipped women or men with six-pack abs? Actually, Geoffrey Miller has argued, the victory goes to the kind.

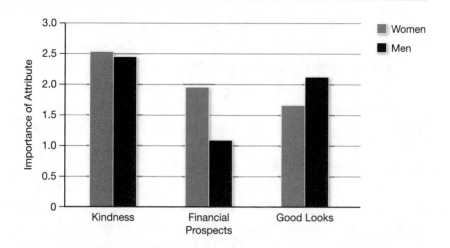

Kindness is the most important quality women and men seek in romantic partners.

Consider the data presented in the figure above, from the largest study of mate preferences ever undertaken, involving 10,000 participants in 37 nations. David Buss asked people at the age of reproduction (20 to 25) to indicate how important different attributes were in potential romantic partners (0 = unimportant, 3 = indispensable). What generated the most heat (and some light) from these findings were the gender differences in mate preferences that remain to this day some of the most highly contested findings in the social sciences: Men prioritize beauty more than women, looking for hourglass-shaped women at the peak of their reproductive potential (see the two bars to the right); women, facing the extreme costs of raising offspring, show a greater preference for silver-haired, ambitious mates with big pocketbooks (see the two middle bars).

Lost in the controversy of this study is another finding: The most important criterion for females and males alike in their search for love, an overwhelming universal across the thirty-seven countries surveyed, is kindness. There are many clear benefits to mating with caretaking individuals, the vagal superstars of our world. They are likely to devote more resources to offspring. They are more likely to provide physical care—touch, protection, play, affection—and create cooperative, caring communities vital to survival. They are more likely to raise offspring that themselves do well in the mating game when they reach the age of reproduction. And presumably, they should be less likely to run off with the next cute thing. The sexual preference for kind individuals makes evolutionary sense, as Darwin long ago surmised: "Sympathy . . . will have been increased through natural selection; for those communities which include the greatest number of the most sympathetic members, would flourish best, and rear the greatest number of offspring."

Evolution went one step further. Social selection pressures—who we favor with friendships, attention, and status in groups—created additional pressure for kindness to be wired into our genes. We only survive socially, in groups, and groups fare better when comprised of kind individuals. In my research, we have asked individuals in different groups to talk in free form about the reputations of randomly selected group members. We provided an opportunity for sorority sisters to gossip about each other by simply asking them to tell nicknames of other sorority sisters not present and what kinds of activities justified those nicknames. The central issue in this kind of reputational discourse was not what you might think—the group member's tendency to drink too much or take illegal drugs, or irritating idiosyncracies (the tendency to play the drums at 2:00 AM, to not do the dishes, or to leave dirty socks or underwear out for all to view and smell). Instead, the central focus of reputational discourse is the kindness and warmth of other group members. Off-the-record chat, banter, and gossip all center upon who lacks kindness and compassion and poses a threat to the harmony of the group. We ferret out cold, self-interested, backstabbing Machiavellians through reputational processes—gossip, casual conversations about the latest things other group members have done.

In fact, so important was the capacity to care to the survival of our species that new data suggest that we have been wired to identify the trustworthy and reliable caretakers among us, and preferentially trust, and give resources to, those vagal superstars. In a study that explored this reasoning, participants played the trust game with a set of vagal superstars and a set of low-vagal-tone individuals, whom we'll call Machiavellians. These participants first viewed

each vagal superstar or Machiavellian for twenty seconds on videotape in a conversation with another person. The sound was off. The
cues our vagal superstars and Machiavellians were giving off were
minimal (a few head nods, an open-handed posture, a gleam in the
eye). The task for our participants was to indicate how much they
trusted each vagal superstar or Machiavellian. They then gave some
amount of money to each vagal superstar and Machiavellian, which
would be sent to that person over the Internet and tripled. That
individual on videotape would then give some amount back to our
participants.

As in life, the task for our new participants was to trust the right
people. Gifts to the more cooperative vagal superstars would more
likely be returned in kind. Avoiding generosity toward the Machiavellians would prevent the participants from being exploited by
these competitive types. And indeed, our new participants trusted
the vagal superstars more. They also gave them more money. The
branches of the nervous system that support compassion and altruism are detected and rewarded in brief encounters with strangers.
It pays to be kind.

SYMPATHY BREAKTHROUGH

When asked what unites the ethics of the world's religions, scholar
Karen Armstrong responded with the simplest of answers: "compassion." If faced with their own version of the question—What is
the central moral adaptation produced in the evolution of human
sociality?—evolutionists would converge on a similar answer: "compassion." On this, the religiously inclined and evolutionists would
agree.

The centrality of compassion to cooperative, high *jen* communities makes it a ready, and necessary, target of attack by those with
contrasting visions of human social life. Hitler knew that compassion—sympathy breakthroughs—could undermine his master plan:

My pedagogy is hard. What is weak must be hammered away.
In my fortresses of the Teutonic Order a young generation will

grow up before which the world will tremble. I want the young to be violent, domineering, undismayed, cruel. The young must be all these things. They must be able to bear pain. There must be nothing weak or gentle about them. The free, splendid beast of prey must once again flash from their eyes.

Early practices of Hitler's SS—shooting women and children in face-to-face encounters—led to drinking, depression, and desertion. As a result, the training of the SS officers shifted, in ways that hammered out of the soul all that was gentle, leaving only that flash in the eyes of the predator. SS officers were ordered to use Jews for target practice. Some SS officers were asked to kill their pets with their own hands. Jews were dehumanized, treated as animals in cattle cars, made to defecate in public, and used in scientific experiments on the limits of pain.

Today, we are engaged in a more subtle struggle over compassion. It is not found in a demagogue's ideology or Fascist's social engineering but in the content of our culture. Violent video games, ad-filled Internet sites, and the new digital world of "weak ties" all diminish the face-to-face and skin-to-skin basis of compassion. This struggle is likely shaping the nervous systems of our children, perhaps in permanent ways. Recent neuroscientific evidence suggests that the regions of the brain that enable compassion—portions of the frontal lobes involved in empathy and perspective taking—continue to develop into the twenties. Compassion can be cultivated.

When Richie Davidson scanned the brain of a Tibetan monk, he found it to be off the charts in terms of its resting activation in the left frontal lobes. This region of the brain supports compassion-related action, feeling, and ideation. After years of devotion and discipline, his was a different brain, humming with compassion-related neural communication.

Okay, you're rightfully critiquing, whose resting brain state wouldn't shift to the left if you had the time and steadfastness to meditate for four to five hours a day upon loving kindness, as Tibetan Buddhists do? Fair enough. When Richie and Jon Kabat-Zinn and colleagues had software engineers train in the techniques of mindfulness meditation—an accepting awareness of the mind, loving kindness toward others—six weeks later these individuals showed increased activation in the left frontal lobes. They also showed enhanced immune function. They may not have been donning the saffron robes of the monk, but at least their minds were moving in that kind direction.

Recent scientific studies are identifying the kinds of environments that cultivate compassion. This moral emotion is cultivated in environments where parents are responsive, and play, and touch their children. So does an empathic style that prompts the child to reason about harm. So do chores, as well as the presence of grandparents. Making compassion a motif in dinnertime conversations and bedtime stories cultivates this all-important emotion. Even visually presented concepts like "hug" and "love" at speeds so fast participants couldn't report what they had seen increase compassion and generosity.

Compassion is that powerful an idea. It is a strong emotion, attuned to those in need. It is a progenitor of courageous acts. It is wired into our nervous systems and encoded in our genes. It is good for your children, your health, and, recent studies suggest, it is vital to your marriage. In the words of the Dalai Lama: "If you want to be happy, practice compassion; if you want others to be happy, practice compassion." It has taken a sympathy breakthrough for science to catch up to this wisdom of the ages. Ironically enough, compassion may be a prerequisite to the pursuit of self-interested happiness.

12

Awe

ONE AFTERNOON in a botany class at the University of Wisconsin, Madison, John Muir heard a fellow student explain how the flower of an enormous black locust tree is a member of the pea family. That the giant black locust tree and the frail pea plant, so remote in size, form, and apparent design, shared an evolutionary history astounded Muir. He later wrote: "This fine lesson charmed me and sent me flying to the woods and meadows in wild enthusiasm."

Shortly thereafter, Muir left college. He walked 1,000 miles on a naturalist's pilgrimage to Florida. He then moved west, to California, and in the summer of 1869, at the age of twenty, herded a couple hundred sheep through the Sierra Nevada Mountains on a trail that wound its way to Yosemite. During this trip he kept a small diary attached to his leather belt. He wrote almost daily entries about these first experiences, which eventually were published as *My First Summer in the Sierras*. A few days into this trip, Muir writes:

June 5
a magnificent section of the Merced Valley at what is called Horseshoe Bend came full in sight—a glorious wilderness that seemed to be calling with a thousand songful voices. Bold,

down-sweeping slopes, feathered with pines and clumps of manzanita with sunny, open spaces between them, make up most of the foreground; the middle and background present fold beyond fold of finely modeled hills and ridges rising into mountain-like masses in the distance. . . . The whole landscape showed design, like man's noblest sculptures. How wonderful the power of its beauty! Gazing awestricken, I might have left everything for it. Glad, endless work would then be mine tracing the forces that have brought forth its features, its rocks and plants and animals and glorious weather. Beauty beyond thought everywhere, beneath, above, made and being made forever.

The next day Muir's immersion in the boundless beauty of the Sierras yielded the following:

June 6
We are now in the mountains and they are in us, kindling enthusiasm, making every nerve quiver, filling every pore and cell of us. Our flesh-and-bone tabernacle seems transparent as glass to the beauty about us, as if truly an inseparable part of it, thrilling with the air and trees, streams, and rocks, in the waves of the sun—a part of all nature, neither old nor young, sick nor well, but immortal. . . . How glorious a conversion, so complete and wholesome it is, scare memory enough of the old bondage days left as a standpoint to view it from!

Muir's experiences in the Sierras opened his mind to new scientific insights: He was the first to argue that Yosemite Valley was formed by glaciers, as opposed to earthquakes, the conventional wisdom of the day. Out of these experiences Muir published on the need to preserve the Sierras from the ravages of sheep and cows in the influential magazine *Century*. These well-placed essays led to a bill passed by Congress on September 30, 1890, designating Yosemite as a state park. Buoyed by this success, Muir founded the Sierra club in 1892 and served as its first president until his dying day.

Today, when back-country hikers find high-altitude *jen* on the John Muir Trail in the Sierras, they are there because of John Muir. So too are groups of inner-city children backpacking near Yosemite in programs sponsored by the Sierra Club. When psychologist Frances Kuo finds in her research that adding trees and lawns to housing projects in Chicago leads local residents to feel greater calm, focus, and well-being, and crime rates drop, she is testing hypotheses that trace back to Muir's transformative experiences of awe.

The thread that awe weaves through the life of John Muir is as revealing about the structure of this transcendent emotion as any study a scientist might deign to conduct. It is a high-wattage experience, nearly as rare as birth, marriage, and death, one that transforms people, energizes them in the pursuit of the meaningful life and in the service of the greater good. Science, until recently, has shied away from the study of awe. Perhaps Lao Tzu's admonition is right:

> *The way that can be spoken of*
> *Is not the constant way;*
> *The name that can be named*
> *Is not the constant name.*
> *The nameless was the beginning of heaven and earth;*
> *The named was the mother of the myriad creatures.*

Perhaps science, built upon essentialist names and quantification, could never unearth the secrets of awe. Perhaps matters of the spirit operate according to different laws than materialistic conceptions of human nature. Not to be deterred by these concerns, evolutionists have recently begun to make the case that Muir's experiences of wonder and awe are examples of emotions designed to enable people to fold cooperatively into complex social groups, to quiet the voice of self-interest, and to feel a sense of reverence for the collective.

A BRIEF HISTORY OF AWE

That John Muir could stand in the Sierras and experience a sense of the sacred when surrounded by the pine, manzanita, granite, cascading water, and dark lakes of those mountains is a testimony to radical thinkers who fought pitched battles about the nature of the sublime (awe) and the beautiful. These thinkers liberated the experience of awe, wonder, and the sacred from the strictures of organized religion, which had laid claim to this powerful emotion, no doubt because of its transformative powers. Most directly, Muir's experience in the Sierras traces back to Ralph Waldo Emerson:

> In the woods, we return to reason and faith. There I feel that nothing can befall me in life—no disgrace, no calamity (leaving me my eyes), which nature cannot repair. Standing on the bare ground,—my head bathed by the blithe air, and uplifted into infinite space,—all mean egotism vanishes. I become a transparent eye-ball; I am nothing; I see all; the currents of the Universal Being circulate through me; I am part or particle of God. The name of the nearest friend sounds then foreign and accidental: to be brothers, to be acquaintances, master or servant, is then a trifle and a disturbance. I am the lover of uncontained and immortal beauty.

And Emerson could preach transcendentalism in nature as a result of Enlightenment philosophers, in particular Edmund Burke, whose more secular musings provide clues to how our capacity for awe and wonder evolved.

Early in human history awe was reserved for feelings toward divine beings. Paul's conversion on the road to Damascus involved a blinding light, feelings of awe and terror, and a voice guiding him to abandon his persecution of the Christians. In the climax of the Hindu *Bhagavad Gita*, the hero of the story, Arjuna, loses his nerve on the eve of battle. To provide Arjuna a sense of higher purpose, Krishna (a form of the god Vishnu) gives Arjuna a "cosmic eye" allowing Arjuna to see gods and suns and to experience infinite time and space. He is filled with amazement (*vismitas*). His hair stands on end. He prostrates himself before Krishna, begs for forbearance, and hears and heeds Krishna's command: "Do works for Me, make Me your highest goal, be loyal-in-love to Me, cut off all [other] attachments . . ."

In 1757, with the age of enlightenment, political revolution, and the promise of science in the air, Irish philosopher Edmund Burke transformed our understanding of awe. In *A Philosophical Enquiry into the Origin of Our Ideas of the Sublime and Beautiful*, Burke detailed how we feel the sublime (awe) in hearing thunder, in viewing art, in hearing a symphony, in seeing repetitive patterns of light and dark, even in response to certain animals (the ox) rather than others (a cow). Odors, Burke observed, could not produce the feeling of the sublime. In these mundane and purely descriptive observations, Burke was advancing a radical claim fitting for his times: Awe is not restricted to experiences of the divine; it is an emotion of expanded thought and greatness of mind that is produced by literature, poetry, painting, viewing landscapes, and a variety of everyday perceptual experiences.

Burke believed the two essential ingredients to the experience of awe are power and obscurity. On power, Burke wrote: "wheresoever we find strength, and in what light soever we look upon power we shall all along observe the sublime the concomitant of terror." On obscurity, Burke argued that awe follows from the perception of objects that the mind has difficulty grasping. Obscure images in painting are more likely to produce sublime feelings (Monet) than those that are clearly rendered (Pissarro). Despotic governments keep their leader obscure from the populace to enhance that leader's capacity to evoke awe.

Today in the West, awe has been liberated; we are following in Burke's footsteps. In my research, when I ask individuals to recount their last experience of awe, they most typically recall experiences of interest to Burke. They write about nature, art, charismatic, famous people, experiences of the sacred, powerful perceptual experiences, experiences when meditating or praying or contemplating the divine. But the spirit of democracy has spread through awe. People are also likely to recall experiences of awe when the Red Sox broke the curse, when hearing Steve Reich for the first time, after a bowl of celery soup at Chez Panisse, at the end of *The Brothers Karamazov*, when lifted aloft in the mosh pit at an Iggy Pop show, when an insight about their past occurs during therapy, at the birth of their children, their last experience of sex, drinking wine, a trip on LSD, a lucid dream. Awe has been used in the service of unadulterated evil—one only needs to think of Hitler's rallies to realize how readily this sacred emotion can be used to malevolent ends.

To bring some order to this cacophony of transcendence, Jon Haidt and I offered the following analysis of the varieties of awe (see table below).

AN APPROACH TO AWE AND ITS RELATED STATES

	CENTRAL FEATURES		PERIPHERAL OR "FLAVORING" FEATURES				
	VASTNESS	ACCOMMODATION	THREAT	BEAUTY	ABILITY	VIRTUE	SUPERNATURAL
ELICITING SITUATIONS							
SOCIAL ELICITORS							
1. POWERFUL LEADER	X	X	?				
2. ENCOUNTER WITH GOD	X	X	?	?		X	X
3. GREAT SKILL (ADMIRATION*)		X			X		
4. GREAT VIRTUE (ELEVATION*)		X				X	
PHYSICAL ELICITORS							
5. TORNADO	X	X	X	?			?
6. GRAND VISTA	X	X		X			
7. CATHEDRAL	X	X		X	X		?
8. AWE-INSPIRING MUSIC	X	X		X	X		
COGNITIVE ELICITORS							
9. GRAND THEORY	X	X		?			
10. EPIPHANY	X	X					

X DENOTES THAT THE APPRAISAL IS USUALLY MADE IN THIS CASE

? DENOTES THAT THE APPRAISAL IS SOMETIMES MADE IN THIS CASE (AND IF IT IS MADE, IT ADDS A FLAVOR)

* DENOTES STATES THAT ARE RELATED TO AWE BUT SHOULD NOT BE LABELED AS AWE

Prototypical experiences of awe involve perceived vastness, anything that is experienced as being much larger than the self or the self's typical frame of reference. Vastness can be physical (standing next to a 389-foot redwood, seeing Shaquille O'Neal's size 22 high-tops or the expanse of Chichén Itzá). Vastness can be acoustic (thunder, a thunderous electric organ). Vastness can be social (standing near the Dalai Lama, dining next to a celebrity). Ideas, feelings, and sensations can be vast when they transcend what has been known or felt before. Vastness becomes awe-inspiring when it requires accommodation—the process by which we update and change our core beliefs.

The spine-tingling, jaw-dropping experiences of awe involve vastness and accommodation. Our experiences of powerful, charismatic humans, our experiences in nature—when viewing mountains, vistas, storms, redwoods, oceans, tornadoes, earthquakes; our experiences of astonishing artifacts—cathedrals, skyscrapers, sculptures, fireworks, the world's largest ball of string; our feeling of awe when immersed in the breadth and scope of a grand theory (feminism, Marxism, evolutionary theory) . . . are all founded on the sense of vastness and transcendence of our understanding of the world.

The varieties and nuances of awe derive from additional flavoring themes (see columns 4 through 8 in the table). The sense of threat gives rise to awe experiences that have elements of fear; charismatic leaders (Hitler, versus Gandi) or natural scenes (an electrical storm, versus a sunset) evoke awe-related experiences that can feel dangerous or reassuring. Aesthetic properties of the stimulus (its harmony, balance, and proportionality), color awe experiences with the feeling of beauty (in hearing a symphony; viewing the mirror image of a mountain in a lake). Encounters with people of exceptional ability will trigger a related state, admiration. Encounters with extraordinary virtue will trigger the feeling of elevation, an emotional response to "moral beauty" or human goodness. Admiration and elevation are closely related to awe but typically do not involve perceived vastness or power. When supernatural ideation suffuses the experience of awe—the felt presence of nonmaterial entities such as spirits, or supernatural causal processes—the experience of awe acquires a religious flavor. Epiphanies feel awesome because they involve seemingly trivial, incidental events that reveal unexpected, vast truths: A falling leaf reminds you of your father's death, and of your own mortality; a subtle lip pucker evident in your beloved directed toward your friend tips you off to a long-suspected secret affair.

The etymological history of the word "awe" parallels this liberation of the experience. "Awe" derived from related words in Old English and Old Norse that were used to express fear and dread, particularly toward a divine being. Now "awe" connotes "dread mingled with veneration, reverential or respectful fear; the attitude of a mind subdued to profound reverence in the presence of supreme authority, moral greatness or sublimity, or mysterious sacredness" (*Oxford English Dictionary*). The state has been transformed from one that centered upon fear and dread to one of reverence, devotion, and pleasure.

IN THE BEGINNING

The Greek philosopher Protagoras, source of the saying "Man is the measure of all things," offered the following myth about human origins. For some time, only gods existed on Earth. The gods decided to create the different species, not out of a primordial molecular soup but out of earth and fire. The gods distributed the various capacities and abilities—speed, strength, thick hides, tough hooves, agility, tastes for roots or grasses or meat—to the different species so that they would each occupy specific niches and thrive in their own particular ways.

The gods ran out of abilities and talents, alas, before figuring out what to do with that thin-skinned, slow-footed species—humans—who were scattered about in semi-functioning, soon to be extinct bands. Reacting to this state of affairs, Prometheus gave the first humans technology—fire. Zeus, however, quickly realized the limitations of technology. Fire could provide warmth, a means of burning germs out of meat, and forms of defense, but humans would need more to survive; they would need to be bound together in cooperative, strong communities. So Zeus gave humans two qualities. The first is a sense of justice, to ensure that the needs of all would be met. The second was reverence, or the capacity for awe.

In his beautifully distilled book *Reverence*, philosopher Paul

Woodruff reveals in his analysis of ancient Greek and Chinese cultures why our capacity for awe ranks so high on Zeus's list of prerequisites for the prospects of an enduring human culture (I risk offense by summarizing his argument in the accompanying flow chart).

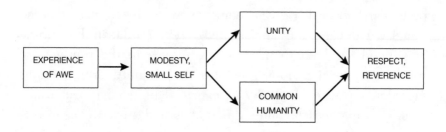

Awe and reverence

Awe is triggered by experiences with that which is beyond our control and understanding—that which is vast and requires accommodation. This experience, at its core, centers upon the recognition of the limitations of the self; in Confucian thought, we feel a deep sense of modesty. Around the world awe has a modest physical signature seen in acts of reverence, devotion, and gratitude: we become small, we kneel, bow, relax and round our shoulders, curl into a small, fetal ball (see Darwin's observations on devotion in table below).

ADMIRATION	EYES OPENED, EYEBROWS RAISED, EYES BRIGHT, SMILE
ASTONISHMENT	EYES OPEN, MOUTH OPEN, EYEBROWS RAISED, HANDS PLACED OVER MOUTH
DEVOTION (REVERENCE)	FACE UPWARDS, EYELIDS UPTURNED, FAINTING, PUPILS UPWARDS AND INWARDS, HUMBLING KNEELING POSTURE, HANDS UPTURNED

Darwin's observations of emotions related to awe

Modesty involves placing the self within a larger context. Experiences of awe reveal us to be small iterations of the patterned history of a family or community, small specks of time and matter in the vastness of the universe. Ambitions and crises, desires and longings, are fleeting instants of time. Our culture is a blip in the millions of years of mammalian evolution.

Reverence, Woodruff continues, is grounded in a sense of unity and a feeling of common humanity. For John Muir, the "flesh-and-bone tabernacle" of the self merged with the trees, air, wind, and rock of the Sierras. The perceptual world of discrete objects and forces vanishes; the flimsy screen of rational consciousness, in William James's terms, is lifted. The mind, like a darkened lake illuminated by the light from the movement of a cloud, reveals forces that interconnect and unite—Emerson's "currents of the Universal Being." All objects are animated by the same pattern of vibration of molecules. The structure of the human face reveals the genome that makes up all humans. Mathematical patterns of design unite the life-forms of a tidepool or floor of a forest. Old traditions—Thanksgiving dinner, weddings, toasts, fathers dancing with daughters—fold individuals into time-honored, cooperative patterns of exchange. Out of this perceptual unity emerges a deep sense of common humanity: We were all infants, we all have families, we all experience grief, and laugh; we all suffer; we all die.

And in the end, awe produces a state of reverence, a feeling of respect and gratitude for the things that are given. Rituals build upon this feeling of reverence—we revere birth, we give thanks for food, we honor those who marry, we pay homage to the dead. We bow our head in appreciation of the kindness of strangers and everyday generosity.

Evolutionists like David Sloan Wilson have arrived at their own story about the evolution of awe, which would not seem foreign to Protagoras or Confucius were they studying evolutionary thought today. This thinking assumes that for groups to work well, and for humans to survive and reproduce, we must often subordinate self-interest in the service of the collective. The collective must often supersede the concerns, needs, and demands of the self. Awe evolved to meet this demand of human sociality.

In our hominid predecessors awe first began to occur in the emotional dynamics of collective action—for example in collective defense, in coordinated hunting, in the rapid response to storms, in the mobilization required at the sound of a herd. In these kinds of collective actions, early hominids felt surges of physical power and connection to their kith and kin. Their body movements became synchronized with others, giving rise to the percept that some force coordinates the many, a sense of unifying common purpose, and a fading of the awareness of boundaries between self and other.

These experiences laid down a readiness to respond to all that unites the members of a group, an attunement as potent as our sensitivity to threat or harm or to the vulnerability of a child. The early hominid mind was ready to respond with awe to individuals who unite the collective—highly ornate leaders, dead family members, neonates. The same came to be true for ideas and objects that bring people together in common feeling or action: mythological stories about the origins of people, chant, celebratory dance, the appreciation of cave paintings. Awash in this experience, our hominid predecessors felt small, a sense of restraint, and a sense of commonality and unity with other group members. This capacity for awe, to be moved by that which unites us into collectives, was to be wired into our minds and bodies. It was to become a dynamic force in culture—source of religion, art, sport, and political movements. The scientific study of awe was to do only modest justice to these claims.

FRACTALS, GOOSE BUMPS, AND *T. REX*

Some emotions are absurdly easy to study in the lab. Embarrassment is one: The minute an individual walks into a lab, aware of being analyzed, experimented upon, videotaped, coded by teams of undergraduates working late into the night, and turned into data, that blush begins to wash over the face.

Other emotions are not so easy. At the top of that list is awe, a humbling object of inquiry. Awe requires vast objects—vistas, encounters with famous people, charismatic leaders, 1,000-foot-tall skyscrapers, cathedrals, supernatural events—that don't fit well in the fluorescent-lighted 9' × 12' space of a lab room. Awe requires unexpected, extraordinarily rare events that exceed our current understanding of the world—the birth of a child, the death of a parent, that one time you were in the hotel lobby near Mick Jagger, that freak tornado that ripped down your street during a summer storm, the first time you went to a rock concert, political rally, saw mountain peaks, had sex, ate chocolate ice cream, drank wine in a Parisian café.

The scientific study of awe represents a Zen-like challenge—measuring that which might transcend measurement, planning what can only be unexpected, capturing what is beyond description. But this didn't prevent my students from producing an outpouring of ideas about how to study awe at a weekly lab meeting devoted to the topic. Capture people's stream of consciousness as they stand at the lip of the Grand Canyon, which William James found to be like one animated organism unified in design. Have participants play a cooperation game with the Dalai Lama or, barring that, the seven-foot center for the basketball team. Bring the world's biggest ball of string into the lab and have participants sit next to it. Fill a bus with participants and drive five hours from Berkeley to the Humboldt Redwoods State Park, where the subjects could walk amid the tallest redwoods in the world. Record the instant the oarsmen of the Berkeley crew are so in sync that their selves dissolve and they let out an exultant roar (an experience of awe recounted to me by their coach, Steve Gladstone). We thought of staging epiphanies in the lab, to capture James Joyce's notion of "the significance of trivial things." Have a participant fall into a conversation with a stranger (actually a confederate) in which they discover that two of their parents almost married, and they almost became siblings. Stage a supernatural event—a voice that sounds like their mother, a vision of a ghost, ooze coming out of the walls. As word spread that we were trying to study awe, I was approached by an all-night dance society—a drug-free raver community—about a possible study of the state they descended into during their parties, which they held in an old church.

Recognizing the impracticalities of these studies, I started with words and images, experiences that, having been raised by a litera-ture professor and a painter, are near and dear to my heart. One stu-dent, a devotee of haiku, had undergraduates fill their minds for half an hour with the best of that poetry—flesh-tingling and inspir-ing for him—and look at whether the experience filled them with a sense of unity and common humanity. No such effect. The students weren't quite sure why they were reading this obscure poetry in a windowless psychology lab.

So I turned to images, on the supposition that awe may more uniformly be triggered by the visual modality. Small groups of undergraduates watched images of endlessly unfolding fractals on a forty-eight-inch screen for half an hour, on the assumption that this experience would lead to expansive, communal conversations afterward. When I looked in on the experiment from my lab's control room (from which you can watch, via video feed, participants in rooms nearby), I saw my honors student running the study, clearly just back from Burning Man, with glitter speckling her cheekbones. Groups of what appeared to be electrical engineering and molecular and cell biology students sat, bemused, watching the fractals, deriving mathematical functions that would explain such organic forms. I swear I heard someone mumble, "Didn't Timothy Leary get his PhD from Berkeley?" (He did.)

But the science of awe, notwithstanding these initial missteps, is inching forward. Let's start with where William James started: The autonomic nervous system. In one study we asked people to describe physical sensations that accompanied different positive emotions, including awe. We found that goose bumps are fairly unique to awe (see graph below).

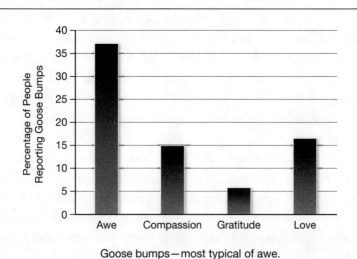

Goose bumps—most typical of awe.

Goose bumps is the colloquial term for piloerection, the activation of minute muscles that surround hair follicles distributed throughout the body but in particular in the back of the neck and back. Piloerection is one action of the fight/flight, sympathetic autonomic nervous system. In our primate relatives—the great apes—piloerection is resorted to in adversarial encounters; primates piloerect to expand their size (with hair standing on end) to threaten and display physical dominance and power. In humans, piloerection shifted in its use, coming to occur regularly when we ourselves feel expanded beyond the boundaries of our skin, and feel connected to other group members. We feel goose bumps when listening to an elevating symphony, when chanting in common cause at a political rally, when hearing a brilliant, mind-expanding lecture, because our self is expanding beyond our physical boundaries to fold into a collective. Piloerection shifted from an association with adversarial defense to connection to the collective.

Alongside piloerection, in the depths of awe people report an expansive, warm swelling in the chest, no doubt a representation of the activation of the vagus nerve. Chris Oveis has found that the vagus nerve does indeed fire during the experience of elevation at others' moral goodness, a close relative of awe. When participants viewed a film about Mother Teresa's works with the poor and starving in Calcutta, their vagus nerve was activated. Awe in the body, then, reflects a confluence of two physiological processes fitting for our evolutionist claims about this transcendent emotion: the expansion of the self in goose bumps, and the opening of the chest to social connection.

This physiological state of awe is accompanied by profound shifts in the sense of the individual's place in the world. In one study, Lani Shiota and I had participants recall transformative experiences in nature, for example when listening to the waves of the Pacific Ocean or walking through the light of a eucalyptus grove. The defining realizations that accompanied these recollections, although lacking the poetic metaphor of Muir or Emerson, were: "I felt small or insignificant," "I felt the presence of something greater than myself," "I felt connected with the world around me," "I was unaware of my day-to-day concerns." Awe diminishes the press of self-interest and reorients the mind to interconnection and design.

Of course, these findings are retrospective, and may just reflect people's theories about what awe does to the mind, rather than what awe actually does to the way that we look at the world. This led Lani to an imaginative study of *in vivo* awe. In this experiment participants arrived at our lab but were told they were to complete the experiment in a different building on campus. They walked for about five minutes across some rolling lawns and a bridge over Strawberry Creek, which winds its way through the Berkeley campus, and arrived at the neoclassical Valley Life Sciences building. They proceeded into the main foyer of the building, where they were asked to sit, not coincidentally, next to a full-sized replica of a *Tyrannosaurus rex* skeleton (see picture below). The skeleton is about twelve feet high at the hip, about twenty-five feet long, and weighs approximately five tons—a source of awe for evolutionists and creationists alike (and in fact, when we stopped other students walking by the *T. rex* and asked them how it made them feel, they put their cell phones aside and consistently uttered "awe").

We then had our participants complete a well-used measure of the self-concept, known as the twenty statements test (TST). In this exercise, participants completed twenty statements beginning with: "I am _____." Control participants completed the same measure, sitting in the same climate-controlled, naturally lit room. Instead of having the *T. rex* looming in their visual field, though, they sat oriented away from it, looking down a hallway. Lani then coded how people described themselves, identifying physical references ("I am redheaded," "I am covered in moles"), trait-based references ("I am gregarious," "I am fragile"), relationship-based references ("I am a nephew," "I am Sherman's main squeeze"), and, a category rarely mentioned but of theoretical interest, an oceanic universal category, where the individual completes the prompt with references to membership in large, social collectives ("I am an organic form," "I am an inhabitant of the Earth," "I am part of the human species"). People feeling awe—that is, those describing themselves while looking at the skeleton of the *T. rex*— were three times more likely to describe themselves in terms of these oceanic, collective categories than those individuals standing

Natalie and Serafina near the *T. rex* in Berkeley's Valley Life Sciences Building.

in the same exact spot but looking away from the awe-inspiring *T. rex*. Awe shifts the sense of self away from characteristics that separate and delineate—idiosyncratic traits and preferences—to facets that unite and highlight common humanity.

Buoyed by these findings, Emiliana Simon-Thomas and I have sought to locate awe in the brain. The conventional neuroscience wisdom is that there is one reward circuit in the brain, which is activated in response to any kind of pleasure, be it money, a massage, a milkshake, hearing an aria, a raise at work, seeing your smiling infant, the touch of a friend, a smooch from a romantic partner, or a view of mountains. All forms of happiness reduce to a single kind of self-interested pleasure. We, of course, would suggest a different hypothesis, one that argues for distinct regions of the brain engaged in different kinds of pleasure and satisfaction. We would expect evolution to have built into the brain different neural circuits that enable the individual to engage in different kinds of positive emotion, be it about taste and smell, or the strength of the self, or about being good to others, or in the presence of that which is vast and beyond our current understanding.

To test this hypothesis, we first culled databases to find slides that elicit sensory pleasure, pride, compassion, and awe. We then had participants view series of these slides while having images of their brains taken in an fMRI scanner. The results strongly suggest that awe, compassion, and pride are not reducible to sensory pleasure; that there is more to good feeling and pleasure than self-interested rewards.

The images of sensory pleasure—hammocks on tropical beaches, pictures of steaming pizza—did just what you would expect from

Sensory pleasure

Pride

Compassion

Awe

the neuroscience literature: They activated the nucleus accumbens, a region of the brain implicated in anticipation and registration of rewarding stimuli, including food and money. The images of sensory pleasure also activated the left dorsal lateral prefrontal cortex and hippocampus—involved in memory and reflective thought (clearly our participants were reflecting, perhaps longingly, upon past sensory pleasures).

The pride slides—images of Berkeley landmarks—activated the rostral medial prefrontal cortex. This region of the frontal lobes has been consistently found to light up when people think about themselves—a perfectly sensible finding, given the self-referential core of pride.

The images of harm and suffering activated bundles of neurons that tell a coherent story about where compassion is in the brain. These images activated the amygdala. These slides also activated a portion of the frontal lobes known as the dorsal medial prefrontal cortex, which is involved in empathy and taking the perspective of another. Compassion integrates the sense of harm and the appreciation of the other's experience.

Finally, awe. The awe slides activated the left orbitofrontal cortex. This region lights up when we are physically touched, and when we anticipate rewards. It is centrally involved in approach and goal-directed action. It is activated in instances in which people reflect upon their own internal experience, from a broader perspective. There are many forms of happiness in the brain; not everything reduces to self-interested pleasure.

A Darwinian study of awe is documenting the physiological underpinnings of our capacity to devote ourselves to the collective. It involves the bodily manifestation of expanding beyond ourselves (goose bumps) and connection (the vagus nerve). It transforms self-representation from that which separates to that which unites. It activates regions of the brain associated with goal-directed behavior and approach, a perspective upon the self, and pleasure. In its ultimate origins in evolution, the sacred is social. Our capacity for wonder and reverence is rooted in the body.

WIRED FOR *JEN*

The experience of awe is about finding your place in the larger scheme of things. It is about quieting the press of self-interest. It is about folding into social collectives. It is about feeling reverential toward participating in some expansive process that unites us all and that ennobles our life's endeavors.

For Charles Darwin, it was his trip on the *Beagle*, and transcendent experiences in the Andes, in the Galápagos, around the cape, and with the hunter-gatherer people he saw during his five-year voyage. After wandering through a forest in the Amazon, he mused: "It creates a feeling of wonder that so much beauty should be apparently created for such little purpose." The forest was "a temple filled with varied productions of the God of Nature." In his observations of flowers, and beetles, and flatworms, and armadillos and trees, he began to discern some force—natural selection—that united them all. Humans were, in the words of Darwin biographer Janet Browne, "just a small part of a much larger interlocking system of life on earth."

For cell biologist Ursula Goodenough, the biochemical processes that make up life and living are sacred. How life-forms emerged in the hot mud of billions of years ago, how two sex cells combine to develop into the human, how DNA evolved over time—these questions stirred her soul. Her understanding of these biological processes is filled with the sense of design, beauty, and vastness that stirred Muir's feelings about the Sierras.

In my short scientific life, my feelings of wonder and reverence began one moment in a lab as a post-doc, a late afternoon when I first began applying the tools of the Facial Action Coding System. These tools allowed me to freeze human action in the millisecond frame of a videotape and take a Darwinian journey, tracing our positive emotions as they manifest today back in evolutionary time to the social dynamics that gave rise to such forms. The emotions that I have been so fortunate to capture in my lab, just for a fragile, fleeting instant, have their evolutionary provenance in a reverence and respect for others, and "identification of ourselves with the beautiful which exists in thought, action, or person, not our own." A teenager's blush triggers a forgiving smile from parents, and conflict and tension subside. A deferential smile and "thank you" between bag boy and elderly woman in the checkout line spread respect and enhance our faith in the human endeavor, if only for a moment or two. Parents, pushing infants on swings, fill a space with smiles, coos, and laughs, creating a warm environment of trust and good-will. Songs of laughter ripple through couples, friends, families, auditoriums, linking minds in cooperative, lighthearted play. With the subtle turn of a phrase or use of the voice, spouses and siblings and parents and their children transform thorny conflicts into play-ful banter. Kind embraces spread from child to friend to grandpar-ent. We have neuropeptides that enable trust and devotion, and a branch of nerves that connects the brain, the voice, and the heart that enables caretaking. Our capacity for awe has given us art, a sense of the sacred. We have genes, neurotransmitters, and regions of the brain that serve these emotions as we serve others. These emotions are the substance of *jen*. Evolution has produced a mind that evolves toward an appreciation of the vastness of our collective design, and emotions that enable us to enact these loftier notions. We are wired for good.

NOTES

3   in honor of the Confucian concept of *jen*: For an excellent summary of Confucius's life, times, and philosophy, see Karen Armstrong, *The Great Transformation* (New York: Anchor, 2006), 240–51.

3   *Jen* is the central idea in: Wing-Tsit Chan, trans., *A Source Book in Chinese Philosophy* (Princeton, NJ: Princeton University Press, 1963), chap. 6. Below are some other quotes from Wing-Tsit Chan's translation of the analects of Confucius that provide an understanding of *jen*:

1:2   Few of those who are filial sons and respectful brothers will show disrespect to superiors, and there has never been a man who is not disrespectful to superiors and yet creates disorder. A superior man is devoted to the fundamentals (the root). When the root is firmly established, the moral law (Tao) will grow. Filial piety and *brotherly* respect are the root of humanity (*jen*).

1:3   A man with clever words and an ingratiating appearance is seldom a man of humanity.

1:6   Young men should be filial when at home and respectful to their elders when away from home. They should be earnest and faithful. They should love all extensively and be intimate with men of humanity. When they have any energy to spare after the performance of moral duties, they should use it to study literature and the arts.

3:3   If a man is not humane (*jen*), what has he to do with ceremonies (*li*)? If he is not humane, what has he to do with music?

4:2   One who is not a man of humanity cannot endure adversity for long, nor can he enjoy prosperity for long. The man of humanity is naturally at ease with humanity. The man of wisdom cultivates humanity for its advantage.

4:3   Only the man of humanity knows how to love people and hate people.

4:4   If you set your mind on humanity, you will be free from evil.

4:5   Confucius said, "Wealth and honor are what every man desires. But if they have been obtained in violation of moral principles, they must not be kept. Poverty and humble station are what every man dislikes. But if they can be avoided only in violation of moral principles, they must not be avoided. If a superior man departs from humanity, how can he fulfill that name? A superior man never abandons humanity even for the lapse of a single meal. In moments of haste, he acts according to it. In times of difficulty or confusion, he acts according to it."

4:6   Confucius said, "I have never seen one who really loves humanity or one who really hates inhumanity. One who really loves humanity will not place anything above it. One who really hates inhumanity will practice humanity in such a way that inhumanity will have no chance to get at him. Is there anyone who has devoted his strength to humanity for as long as a single day? I have not seen anyone without sufficient strength to do so. Perhaps there is such a case, but I have never seen it."

6:21   Confucius said, "The man of wisdom delights in water; the man of humanity delights in mountains. The man of wisdom is active; the man of humanity is tranquil. The man of wisdom enjoys happiness; the man of humanity enjoys long life."

6:28   Tzu-kung said, "If a ruler extensively confers benefit on the people and can bring salvation to all, what do you think of him? Would you call him a man of humanity?" Confucius said, "Why only a man of humanity? He is without doubt a sage. Even (sage-emperors) Yao and Shun fell short of it. A man of humanity, wishing to establish his own character, also establishes the character of others, and wishing to be prominent himself, also helps others to be prominent. To be able to judge others by what is near to ourselves may be called the method of realizing humanity."

7:6   Confucius said, "Set your will on the Way. Have a firm grasp on virtue. Rely on humanity. Find recreation in the arts."

7:29   Confucius said, "Is humanity far away? As soon as I want it, there it is right by me."

5   Engaging in five acts of kindness a week: Sonja Lyubomirsky, *The How of Happiness* (New York: Penguin, 2007), 127–28.

5   Spending twenty dollars on someone else: E. W. Dunn, L. B. Akin, and M. I. Norton, "Spending Money on Others Promotes Happiness," *Science* 319, no. 5870, (2008): 1687–88.

5   When pitted against one another in competitive economic games: A. Dreber, D. G. Rand, D. Fudenberg, and M. A. Nowak, "Winners Don't Punish," *Nature* 452 (2008): 348–51.

6   New neuroscience suggests we are wired for *jen*: J. K. Rilling et al., "A

Neural Basis for Social Cooperation," *Neuron* 35 (2002): 395–405.

6   In over twenty studies: T. N. Bradbury and F. D. Fincham, "Attributions in Marriage: Review and Critique," *Psychological Bulletin* 107 (1990): 3–33.

6   see hidden virtues accompanying their partner's foibles and faults: In this research, romantic partners were asked to write about their partner's greatest fault. More satisfied romantic partners were more likely to see virtue in their partner's faults, and to reflexively offer "yes, but" refutations of the fault. See S. L. Murray and J. G. Holmes, "Seeing Virtues in Faults: Negativity and the Transformation of Interpersonal Narratives in Close Relationships," *Journal of Personality and Social Psychology* 65 (1993): 707–23.

6   In 1996, Paul Zak: P. J. Zak, "Trust," *The Capco Institute Journal of Financial Transformation* 7 (2003): 13–21.

7   signs of a loss of *jen* in the United States are incontrovertible: For the most comprehensive assessment of the well-being of our culture, see David G. Myers, *The American Paradox* (New Haven, CT: Yale University Press, 2000).

7   U.S. adults now have one-third fewer close friends: M. McPherson, L. Smith-Lovin, and M. E. Brashears, "Social Isolation in America: Changes in Core Discussion Networks over Two Decades," *American Sociological Review* 71 (2006): 353–75.

7   In a recent UNICEF study of twenty-one industrialized nations: The overall score is based on a sum of six categories: material well-being, health and safety, education, peer and family relationships, behaviors and risks, and children's own subjective well-being. "UNICEF Ranks Well-Being of British, U.S. Children Last in Industrialized World," *USA Today*, February 14, 2007.

8   I see these disheartening social trends as the culmination of a broader ideology about human nature: Barry Schwartz was one of the first scholars to deconstruct the deeper assumptions of self-interest underlying the social and biological sciences. Barry Schwartz, *Battle for Human Nature* (New York: W. W. Norton, 1986). For another excellent challenge of the assumption that humans are self-interested, one more focused on the debate over altruism, see Alfie Kohn, *The Brighter Side of Human Nature* (New York: Basic Books, 1990).

9   When "pleasure centers" were first discovered in 1954: J. Olds and P. Milner, "Positive Reinforcement Produced by Electrical Stimulation of Septal Area and Other Regions of Rat Brain," *Journal of Comparative and Physiological Psychology* 47 (1954): 419–27.

9   And now a new field, neuroeconomics: Neuroeconomics is the scientific study of whether basic principles of behavioral economics—loss aversion, favoring current gains over future ones, risk taking and risk seek-

ing—are represented in different regions of the brain. The study of reward circuits in the brain is one of the hot areas in neuroeconomics. See B. Knutson and J. C. Cooper, "Functional Magnetic Resonance Imaging of Reward Prediction," *Current Opinions in Neurology* 18, no. 4 (2005): 411–17.

9   consider the debate about generous acts toward strangers: For a lucid examination of altruism, consult the work of Daniel Batson. Batson argues that kind actions that enhance the welfare of others, even at costs to the self, are motivated by selfish reasons, like the desire to reduce personal distress and to receive social praise, as well as pure altruistic motives that stem from a concern for the welfare of others. D. C. Batson and L. L. Shaw, "Evidence for Altruism: Toward a Pluralism of Prosocial Motives," *Psychological Inquiry* 2 (1991): 107–22. Elliot Sober and David Sloan Wilson also provide a thoughtful examination of this debate. Sober and Wilson, *Unto Others: The Evolution and Psychology of Unselfish Behavior* (Cambridge, MA: Harvard University Press, 1998). And for an earlier treatment of the philosophical mistakes made in assuming that altruism is necessarily self-interested, see Kohn, *The Brighter Side of Human Nature*.

9   That the bad is stronger than the good is evident in several findings: R. F. Baumeister, E. Bratslavsky, C. Finkenauer, and K. D. Vohs, "Bad Is Stronger Than Good," *Review of General Psychology* 5 (2001): 323–70. S. E. Taylor, "Asymmetrical Effects of Positive and Negative Events: The Mobilization-Minimization Hypothesis," *Psychological Bulletin* 110 (1991): 67–85. P. Rozin and E. B. Royzman, "Negativity Bias, Negativity Dominance, and Contagion," *Personality and Social Psychology Review* 5 (2001): 296–320.

9   Economic losses loom larger than their equivalent gains: A. Tversky and D. Kahneman, "The Framing of Decisions and the Psychology of Choice," *Science* 21 (1981): 453–58.

9   Slides of negative stimuli: T. A. Ito, J. T. Larsen, N. K. Smith, and J. T. Cacioppo, "Negative Information Weighs More Heavily on the Brain: The Negativity Bias in Evaluative Categorizations," *Journal of Personality and Social Psychology* 75 (1998): 887–900.

10   Freud: *On Murder, Mourning and Melancholia*, trans. Michael Huise (New York: Penguin, 2005).

10   Rand: "In the Words of Ayn Rand: Ayn Rand, with Alvin Toffler," *Playboy*, March 1964, 40.

10   Machiavelli: *The Prince*, trans. G. Bull (New York: Penguin, 2003).

10   Williams: *Adaptation and Natural Selection: A Critique of Some Current Evolutionary Thought* (Princeton, NJ: Princeton University Press, 1966), 255.

11   Just ask the parents of children with Williams Syndrome: see www.williams-syndrome.org.

11   This questioning found galvanizing expression: It's hard to overesti-
mate the influence of Robert Frank's superb book on thinking about emotion, morality, and cooperation. In elegant and provocative arguments,
Frank has made the case for the wisdom of the moral emotions, emotions
like gratitude and love, and how these emotions are a bedrock of cooperative communities. Robert H. Frank, *Passions Within Reason* (New York:
W. W. Norton, 1988).

12   economists Ernst Fehr and Klaus Schmidt found that 71 percent of
the allocators offered the responder between 40 and 50 percent of the
money: Fehr and Schmidt, "A Theory of Fairness, Competition, and
Cooperation," *Quarterly Journal of Economics* 114 (1999): 817–68.

12   Does material gain make us happy?: These findings as well as many
others that speak to the rise of materialism in contemporary U.S. culture
are summarized in David Myers, *The American Paradox* (New Haven,
CT: Yale University Press, 2000).

12   Look at the table below, adapted from: Alain de Botton, *Status Anxiety* (New York: Pantheon, 2004).

13   Does money make us happy?: D. G. Myers, "The Funds, Friends,
and Faith of Happy People," *American Psychologist* 55 (2000): 56–67.

13   what makes us happy is the quality of our romantic bonds, the health
of our families, the time we spend with good friends, the connections we
feel to communities: For superb summaries of the many benefits of
healthy relationships, as well as the costs of impoverished relationships
and isolation, see R. F. Baumeister and M. R. Leary, "The Need to Belong:
Desire for Interpersonal Attachments as a Fundamental Human Motivation," *Psychological Bulletin* 117 (1995): 497–529; M. Argyle, "Causes and
Correlates of Happiness," in *Well-Being: The Foundations of Hedonic
Psychology*, ed. Daniel Kahneman, Edward Diener, and Norbert Schwarz
(New York: Russell Sage, 1999), 353–73; Jonathan Haidt, *The Happiness
Hypothesis* (New York: Basic Books, 2006).

13   there are more words in the English language that represent negative than positive emotions: James A. Russell, "Culture and Categorization
of Emotion," *Psychological Bulletin* 110 (1991): 426–50.

14   These empirical facts led many in the field to the view that positive
emotions are in reality by-products of negative states: Sylvan S. Tomkins,
"Affect Theory," in *Approaches to Emotion*, ed. Klaus Scherer and Paul
Ekman (Hillsdale, NJ: Erlbaum, 1984), 163–95.

15   My hope is to tilt your *jen* ratio to what the poet Percy Shelley: Percy
Bysshe Shelley, "A Defence of Poetry," in *The Longman Anthology of
British Literature*, vol. 2 (New York: Longman, 1999).

DARWIN'S JOYS

17   Darwin's: Charles Darwin, *The Expression of the Emotions in Man and Animals*, 3rd ed. (New York: Oxford University Press, 1872/1998).

17   Perhaps most important to Darwin, the book met with modest smiles of approval from his wife, Emma: Janet Browne's brilliant two-volume biography of Darwin reveals the at times ambivalent stance Emma Darwin took with regard to her husband's revolutionary scholarship. Browne, *Charles Darwin: Voyaging*. (Princeton, NJ: Princeton University Press, 1995), and *Charles Darwin: The Power of Place* (Princeton, NJ: Princeton University Press, 2002).

17   anatomist Sir Charles Bell: Browne, *Power of Place*, 364.

23   One of the clearest signs of dominance: For some of the most systematic work on power and nonverbal display, consult the work of John Dovidio. Dovidio et al., "The Relationship of Social Power to Visual Displays of Dominance between Men and Women," *Journal of Personality and Social Psychology* 54 (1988): 233–42. Judith Hall and her colleagues have also provided a definitive review of the different behaviors that accompany displays of power. J. A. Hall, E. J. Coats, and L. S. LeBeau, "Nonverbal Behavior and the Vertical Dimension of Social Relations: A Meta-Analysis," *Psychological Bulletin* 131, no. 6 (2005): 898–924.

23   One prevailing metaphor of emotion: George Lakoff has done superb work on the different metaphors of emotion. George Lakoff and Mark Johnson, *Metaphors We Live By* (Chicago: University of Chicago Press, 1980); Lakoff, *Women, Fire and Dangerous Things: What Categories Reveal About the Mind* (Chicago: University of Chicago Press, 1987). See also Zoltán Kövesces, *Metaphor* (Oxford: Oxford University Press, 2002).

25   Paul Ekman put Darwin's universality thesis to a simple empirical test: The original report of Ekman's study is found in the following: Ekman, E. R. Sorenson, and Wallace V. Friesen, "Pan-Cultural Elements in the Facial Displays of Emotions," *Science* 164 (1969): 86–88. The critiques of this study are best summarized in James Russell's assessments of the data on the universality of emotion recognition in the face: J. A. Russell, "Is There Universal Recognition of Emotion from Facial Expression? A Review of Methods and Studies," *Psychological Bulletin* 115 (1994): 102–41. This critique focuses on several questions. The most important is whether people in different cultures would label Ekman's faces in similar fashion if allowed to use their own words, rather than using the words or scenarios provided by an experimenter. The answer is yes. See J. Haidt and D. Keltner, "Culture and Facial Expression: Open Ended Methods Find More Faces and a Gradient of Universality," *Cognition and Emotion* 13 (1999): 225–66.

25   The data gathered in this study would pit two radically different conceptions of emotion against one another: For an excellent summary of social constructionist accounts of emotion, see Keith Oatley, "Social Construction in Emotion," in *Handbook of Emotions*, ed. Michael Lewis and Jeannette Haviland (New York: Guilford Press, 1993), 342–52.

26   An evolutionary approach took shape as Ekman started to publish the findings from this first study: For an early evolutionary account of the emotions, see John Tooby and Leda Cosmides, "The Past Explains the Present: Emotional Adaptations and the Structure of Ancestral Environments," *Ethology and Sociobiology* 11 (1990): 375–424. R. M. Neese, "Evolutionary Explanations of Emotions," *Human Nature* 1 (1990): 261–83. For a more recent summary of such thinking, see D. Keltner, J. Haidt, and M. N. Shiota, "Social Functionalism and the Evolution of Emotions," in *Evolution and Social Psychology*, ed. Mark Schaller, Jeffrey A. Simpson, and Douglas T. Kenrick (New York: Psychology Press, 2006), 115–42.

26   Emotions at their core are concepts, words, and ideas that shape, and are shaped by, discourse practices such as storytelling, poetry, public shaming, or gossip: Anthropologist Lila Abu-Lughod has done brilliant work on how emotions are embedded in social discourse and constructed in those social practices. In her work on Bedouin culture, she documents how emotions like embarrassment and modesty are constructed in patterns of dress, poetry, and gossip. Lila Abu-Lughod, *Veiled Sentiments* (Berkeley: University of California Press, 1986). Lila Abu-Lughod and Catherine A. Lutz, *Introduction to Language and the Politics of Emotion* (New York: Cambridge University Press, 1990).

27   The Inuit were never observed to express anger: J. L. Briggs, *Never in Anger: Portrait of an Eskimo Family* (Cambridge, MA: Harvard University Press, 1970).

28   In the critical study: Ekman, Sorenson, and Friesen, "Pan-Cultural Elements."

29   Perhaps the chorus of critiques arose because Ekman's data may have been reminiscent of the claims of Social Darwinism: For an enlightening history of Social Darwinism, read Stanford historian Carl Degler's work. Degler, *In Search of Human Nature* (Oxford: Oxford University Press, 1991).

29   Such was the aim of Stanley Schachter and Jerome Singer: Schachter and Singer, "Cognitive, Social and Physiological Determinants of Emotional State," *Psychological Review* 69 (1962): 379–99.

30   E. E. Cummings: *Complete Poems, 1913/1962* (New York: Harcourt, Brace, Jovanovich, 1980).

32   On this, Linda Levine and George Bonanno have found in their research that when people report upon past experiences: L. Levine,

"Reconstructing Memory for Emotions," *Journal of Experimental Psychology: General* 126 (1997): 165–77. M. A. Safer, G. A. Bonanno, and N. P. Field, "It Was Never That Bad: Biased Recall of Grief and Long-Term Adjustment to the Death of a Spouse," *Memory* 9 (2001): 195–204.

32 To capture the objective subjective, Ekman and Wallace Friesen devoted seven years, without funding or promise of publication, to developing the Facial Action Coding System: For a full description of the system, see Ekman and Friesen, *Facial Action Coding System: A Technique for the Measurement of Facial Movement* (Palo Alto, CA: Consulting Psychologists Press, 1978). For a compilation of dozens of studies that have fruitfully applied the Facial Action Coding System to the scientific study of emotion, see Ekman and Erika L. Rosenberg, *What the Face Reveals* (New York: Oxford University Press, 1997).

34 hundreds of studies have discovered that the muscle configurations that Darwin described for many emotions: For summaries of the research on the universality of emotion recognition in the face, see Hilary Elfenbein and Nalini Ambady's definitive reviews. Elfenbein and Ambady, "On the Universality and Cultural Specificity of Emotion Recognition: A Meta-Analysis," *Psychological Bulletin* 128 (2002): 203–35, and "Universals and Cultural Differences in Recognizing Emotions," *Current Directions in Psychological Science* 12 (2003): 159–64.

RATIONAL IRRATIONALITY

36 Had I read Nobel Prize-winning economist: Thomas Schelling, *The Strategy of Conflict* (Cambridge, MA: Harvard University Press, 1963).

36 The commitment problem has two faces: See also Frank, *Passions Within Reason*; R. M. Nesse, "Evolutionary Explanations of Emotions," *Human Nature* 1 (1990): 261–83.

36 The very nature of emotional experience: One of the founding figures in the study of emotion, the Dutch psychologist Nico Frijda, was one of the first to observe that emotions, while being inherently subjective, feel absolute, as if based on non-negotiable truths. Nico H. Frijda, *The Emotions* (Cambridge: Cambridge University Press, 1986), and "The Laws of Emotion," *American Psychologist* 43 (1988): 349–58.

36 The potent pangs of guilt help us repair our dearest relations: Guilt is a source of many pro-social actions. See T. Ketelaar and W. T. Au, "The Effects of Guilty Feelings on the Behavior of Uncooperative Individuals in Repeated Social Bargaining Games: An Affect-as-Information Interpretation of the Role of Emotion in Social Interaction," *Cognition and Emotion* 17 (2002): 429–53. June Tangney and Rowland Miller have also done definitive work on the more pro-social character of guilt, and how it differs from a close relative, shame. J. P. Tangney, "Assessing Individual

Differences in Proneness to Shame and Guilt: Development of the Self-Conscious Affect and Attribution Inventory," *Journal of Personality and Social Psychology* 59, no. 1 (1990): 102–11; "Moral Affect: The Good, the Bad, and the Ugly," *Journal of Personality and Social Psychology* 61 (1991): 598–607; "Situational Determinants of Shame and Guilt in Young Adulthood," *Personality and Social Psychology Bulletin* 18 (1992): 199–206. Tangney, R. S. Miller, L. Flicker, and D. H. Barlow, "Are Shame, Guilt, and Embarrassment Distinct Emotions?" *Journal of Personality and Social Psychology* 70 (1996): 1256–64.

37 Emotional displays provide reliable clues to others' commitments: For an analysis of truthfulness and deception in the signaling of nonhuman species: John R. Krebs and Nicholas B. Davies, *An Introduction to Behavioural Ecology* (Oxford: Blackwell, 1993), chap. 14, and "Animal Signals: Mind Reading and Manipulation," in *Behavioral Ecology*, 2nd ed., ed. Krebs and Davies (Oxford: Blackwell, 1984), 380–402. Paul Ekman has offered precise statements about which muscle movements are the reliable indicators of an accompanying emotional state. See Ekman, "Facial Expression and Emotion," *American Psychologist* 48 (1993): 384–92.

39 leads to actions that enhance the welfare of others: N. Eisenberg et al., "Relation of Sympathy and Distress to Prosocial Behavior: A Multimethod Study," *Journal of Personality and Social Psychology* 57 (1989): 55–66.

39 Robert Frank reasoned, in a synthesis of Schelling's insights and Ekman's methodological labors: Frank, *Passions Within Reason*.

39 It involves the pulling in and upward of the inner eyebrows, and has been shown: N. Eisenberg et al., "Relation of Sympathy and Distress to Prosocial Behavior."

40 Emotions, Martha Nussbaum argues: Nussbaum, *Upheavals of Thought: The Intelligence of Emotions* (New York: Cambridge University Press, 2001), 9–88; See also Abu-Lughod and Lutz, *Introduction to Language and the Politics of Emotion.*

41 It may have been his somatic oversensitivities that led James to publish his radical thesis about emotion: William James, "What Is an Emotion?" *Mind* 9 (1884): 188–205.

41 "My thesis": William James, *The Principles of Psychology*, 2 vols. (Bristol: Thoemmes Press, 1890/1999), vol. 2, 449–50.

41 The most general function of the ANS: For an excellent, detailed overview of the autonomic nervous system, see W. Janig, "The Autonomic Nervous System and Its Coordination by the Brain," in *Handbook of Affective Sciences*, ed. Richard Davidson, Klaus Scherer, and Hill Goldsmith (London: Oxford University Press, 2003), 135–86.

42 The autonomic nervous system is like the old furnace: My colleague

and former mentor Robert Levenson has done some of the most rigorous and original research on James's thesis about emotion. He is the source of the furnace metaphor, and of much of my thinking about the autonomic nervous system's role in emotion. For his own thinking, see Levenson, "Autonomic Specificity and Emotion," in *Handbook of Affective Sciences*, 212–24.

44 What followed was a rather strange and controversial study: Ekman, Levenson, and Friesen, "Autonomic Nervous System Activity Distinguishes Among Emotions," *Science* 221 (1993): 1208–10. For a follow-up study, see Levenson, Ekman, and Friesen, "Voluntary Facial Action Generates Emotion-Specific Autonomic Nervous System Activity," *Psychophysiology* 27 (1993): 363–84.

45 these distinctions are not the kind of emotion-specific physiological signatures that James envisioned: For a superb and fair-minded critique of the work using the Directed Facial Action task, see John T. Cacioppo, D. J. Klein, G. C. Berntson, and E. Hatfield, "The Psychophysiology of Emotion," in *Handbook of Emotions*, ed. Lewis and Haviland, 119–42.

45 Levenson and Ekman subsequently packed their physiological equipment up: Levenson, Ekman, K. Heider, and Friesen, "Emotion and Autonomic Nervous System Activity in the Minangkabau of West Sumatra," *Journal of Personality and Social Psychology* 62 (1992): 972–88.

45 And in other research: Levenson, L. L. Carstensen, Friesen, and Ekman, "Emotion, Physiology, and Expression in Old Age," *Psychology and Aging* 6 (1991): 28–35.

46 "A man goes to the supermarket once a week": For a full complement of these moral scenarios, see J. Haidt, S. H. Koller, and M. G. Dias, "Affect, Culture, and Morality, or Is It Wrong to Eat Your Dog?" *Journal of Personality and Social Psychology* 65 (1993): 613–28.

46 People's responses to this kind of thought experiment have led Jonathan Haidt: Haidt, "The Emotional Dog and Its Rational Tail: A Social Intuitionist Approach to Moral Judgment," *Psychological Review* 108 (2001): 814–34, and "The Moral Emotions," in *Handbook of Affective Sciences*, 852–70. M. D. Hauser, *Moral Minds* (New York: HarperCollins, 2006). J. D. Greene, and Haidt, "How (and Where) Does Moral Judgment Work?" *Trends in Cognitive Sciences* 6 (2002): 517–23.

47 well-known theory of moral development: Lawrence Kohlberg, "Moral Stages and Moralization: The Cognitive-Developmental Approach," *Moral Development and Behavior: Theory, Research, and Social Issues*, ed. T. Lickona (New York: Holt, Rinehart, and Winston, 1976), 31–53. For excellent extensions of some of Kohlberg's claims, see the work of Elliot Turiel: *The Development of Social Knowledge: Morality and Convention.* (Cambridge: Cambridge University Press, 1983); *The Culture of Morality* (Cambridge: Cambridge University Press, 2002);

Turiel, M. Killen, and C. Helwig, "Morality: Its Structure, Functions, and Vagaries," in *The Emergence of Morality in Children*, ed. Jerome Kagan and S. Lamb (Chicago: University of Chicago Press, 1987), 155–243.

47 In one study: D. Keltner, P. C. Ellsworth, and K. Edwards, "Beyond Simple Pessimism: Effects of Sadness and Anger on Social Perception," *Journal of Personality and Social Psychology* 64 (1993): 740–52.

47 consider the following neuroimaging study: J. D. Greene et al., "An fMRI Investigation of Emotional Engagement in Moral Judgment," *Science* 75 (2001): 2105-08.

48 When the Dalai Lama visited: His Holiness the Dalai Lama, *Ethics for the New Millennium* (New York; Riverhead, 1999), 63.

48 Confucius was on the same page: Armstrong, *The Great Transformation*, 249.

48 Martha Nussbaum, bucking the trends of moral philosophy: Nussbaum, *Upheavals of Thought*, 297–454.

50 Emotions have not fared well in these thought experiments: For an excellent survey of treatments of emotion in Western thought, see Keith Oatley, *Emotions: A Brief History* (Malden, MA: Blackwell, 2004).

50 consider the metaphors that we routinely use in the English language to explain our emotions: Lakoff and Johnson, *Metaphors We Live By*; Lakoff, *Women, Fire and Dangerous Things*; Kövesces, *Metaphor*.

SURVIVAL OF THE KINDEST

52 In November 1943, S. L. A. "Slam" Marshall, a U.S. Army lieutenant colonel: S. L. A. Marshall, *Men Against Fire: The Problem of Battle Command* (Norman, OK: University of Oklahoma Press, 1947/2000).

53 they would reach contrasting conclusions: For an excellent summary of early evolutionary views of altruism, see Helena Cronin, *The Ant and the Peacock*. (New York: Cambridge University Press, 1991), chap. 15.

53 In *Descent*, Darwin argued that the social instincts: *The Descent of Man, and Selection in Relation to Sex* (London: John Murray, 1871), chap. 4.

53 "The following proposition": Ibid., 84.

54 "Such actions as the above": Ibid., 95.

55 at the top of my list would be the field notes of a Cro-Magnon anthropologist: Several books portray the social lives of our hominid predecessors: David Christian, *Maps of Time: An Introduction to Big History* (Berkeley, CA: University of California Press, 2004). 139–332; Stephen Mithen, *After the Ice: A Global History of Human History 20,000 to 5000 BC* (Cambridge, MA: Harvard University Press, 2006); Nicholas Wade *Before the Dawn: Recovering the Lost History of Our Ancestors* (New York: Penguin Press, 2007), 51–180.

55 A detailed portrayal of the day in the life of our hominid predecessors would shed light on our environment of evolutionary adaptedness (EEA): John Tooby and Leda Cosmides, "The Psychological Foundations of Culture," in *The Adapted Mind: Evolutionary Psychology and the Generation of Culture*, ed. Jerome H. Barkow, Leda Cosmides, and John Tooby (New York: Oxford University Press, 1992), 163–228.

56 We can turn to studies of our closest primate relatives, chimpanzees and bonobos in particular: Several books capture the social dimensions of mammalian evolution that lay the foundation for the analysis of emotions that I offer: Frans B. M. de Waal, *Good Natured: The Origins of Right and Wrong in Humans and Other Animals* (Cambridge, MA: Harvard University Press, 1996); John R. Krebs and Nicholas B. Davies, *An Introduction to Behavioral Ecology* (Oxford: Blackwell Press, 1993); Jared Diamond, *The Third Chimpanzee* (New York: Harper Perennial, 1992); Frans B. M. de Waal, ed., *Tree of Origin: What Primate Behavior Can Tell Us About Human Social Evolution* (Cambridge, MA: Harvard University Press, 2001), 121–43; Marc Hauser, *The Evolution of Communication*, (Cambridge, MA: MIT Press, 1996).

56 first attempts at visual art and music: Stephen Mithen, *The Prehistory of the Mind: The Cognitive Origins of Art and Science* (London: Thames and Hudson, 1996); D. Christian, *Maps of Time: An Introduction to Big History* (Berkeley, CA: University of California Press, 2004).

56 detailed observations of contemporary hunter-gatherer societies: Paul R. Ehrlich, *Human Natures. Genes, Cultures, and the Human Prospect* (New York: Penguin, 2002); Melvin Konner, *The Tangled Wing: Biological Constraints on the Human Spirit* (New York: Holt, 2003); Irenäus Eibl-Eibesfeldt, *Human Ethology* (New York: Aldine de Gruyter, 1989).

58 the prevalence of caregiving, a hallmark feature of higher primates: Sarah Blaffer Hrdy has written brilliantly about the prevalence of caretaking behavior in different primates and how it is an often overlooked basis of alliances and strategic exchanges. Hrdy, *Mother Nature* (New York: Ballantine, 1999). For a rigorous discussion of human caregiving, read Shelley E. Taylor, *The Tending Instinct* (New York: Henry Holt, 2002).

58 as Frans de Waal has observed: de Waal, *Good Natured: The Origins of Right and Wrong in Humans and Other Animals* (Cambridge, MA: Harvard University Press, 1996), chap. 2.

58 Our hominid predecessors evolved bigger brains: Ehrlich, *Human Natures*, chap. 6.

59 "carried in a sling": Konner, *The Tangled Wing*, 306.

61 Cooperative child rearing, where relatives and friends traded off duties: Hrdy, *Mother Nature*, 90–95.

61 consistent evidence of cooperative hunting for meat: Stephen

Mithen, *Singing Neanderthals: The Origins of Music, Language, Mind, and Body* (Cambridge, MA: Harvard University Press, 2007), 238.

61 fall to their deaths: ibid., 238.

61 morphological changes that gave rise to our remarkable capacity to communicate: Marc D. Hauser, *The Evolution of Communication* (Cambridge, MA: MIT Press, 1996).

61 Unlike our primate relatives, the human face has relatively little obscuring hair: Nina Jablonski, *Skin: A Natural History* (Berkeley, CA: University of California Press, 2006), 43.

61 allowing for a much richer vocabulary of expressive behavior originating in the face: D. Matsumoto et al., "Facial Expressions of Emotion," in *Handbook of Emotion*.

61 The evolving capacity to communicate is even more pronounced in the human voice: Ehrlich, *Human Natures*, 152.

62 to represent and spread information across time and space with language: Robert Boyd and Peter J. Richardson, eds., *Culture and the Evolutionary Process* (Chicago: University of Chicago Press, 1985), 204–40.

62 our basic emotional tendencies can quickly spread to others, through mimicry, imitation, and communication: For a superb review of the evolution of the capacity to imitate and empathize, see S. D. Preston and F. B. M. de Waal, "Empathy: Its Ultimate and Proximate Bases," *Behavioral and Brain Sciences* 25 (2002): 1–72.

62 In research with my colleague Cameron Anderson: C. Anderson, O. P. John, D. Keltner, and A. Kring, "Social Status in Naturalistic Face-to-Face Groups: Effects of Personality and Physical Attractiveness in Men and Women," *Journal of Personality and Social Psychology* 81 (2001): 118–29.

62 Female adults attain comparable levels of status with just as much alacrity and effect: Ibid. We found few differences in how women and men ascribe status to other women and men in their groups. High-status women and men show similar tendencies to keep their elevated positions of power over time. The one difference we did document is that group members achieved consensus, or agreement, in their judgments of the status of men a bit faster than in their judgments of women. It is also important to bear in mind that these are status judgments in informal groups. It is very likely that within institutions with historical sex-based differences in status (for example, the U.S. Senate), one is likely to find differences in judgment.

63 Yet the hierarchical social organization of higher primates and early humans differs dramatically from that of other species: One of the first scientists to make this point was Christopher Boehm. Boehm, *Hierarchy in the Forest: The Evolution of Egalitarian Behavior* (Cambridge, MA: Harvard University Press, 1999). See also Dacher Keltner, "The Power Paradox," *Greater Good* 8 (2008): 14–17.

63   Frans de Waal has found: *Chimpanzee Politics* (New York: Harper and Row, 1982) and *Peacemaking Among Primates* (Cambridge, MA: Harvard University Press, 1989).

63   Instead, it is the socially intelligent individuals who advance the interests of other group members (in the service of their own self-interest) who rise in social hierarchies: Several new lines of research lend support to this counterintuitive claim. In a recent review, we summarize how socially energetic, outgoing individuals gain power in social groups, and how aggressive, manipulative, Machiavellian types often lose power. Andersen et al., "Social Status in Naturalistic Face-to-Face Groups: Effects of Personality and Physical Attractiveness in Men and Women," *Journal of Personality and Social Psychology* 81 (2001): 1108–29. Stephane Coté has recently documented that socially intelligent individuals—that is, those individuals who are able to understand their own emotions and those of other group members—acquire leadership in organizations. S. Coté and C. T. H. Miners, "Emotional Intelligence, Cognitive Intelligence, and Job Performance," *Administrative Science Quarterly* 51 (2006): 1–28. Perhaps the most dramatic example of this thesis is recent work by Cameron Anderson and colleagues, who have found that individuals in social groups who have modest assessments of their own power actually keep positions of power, whereas individuals who have inflated assessments of power lose power over time. C. Anderson et al., "Knowing your Place: Self-Perceptions of Status in Face-to-Face Groups," *Journal of Personality and Social Psychology* 91, no. 6 (2006): 1094–1110. We have long been guided by Machiavellian analyses of power, which suggest that effective leadership requires deception, strategic manipulation, pitting group members against one another, and being feared. The new empirical science of power is proving this to be an erroneous set of assumptions. See Keltner, "The Power Paradox."

64   who can tell a good joke or tease in ways: In our research on teasing we have found that individuals who are highly esteemed by other group members are quite adept at teasing in ways that are playful, D. Keltner et al., "Teasing in Hierarchical and Intimate Relations," *Journal of Personality and Social Psychology* 75, no. 5 (1998): 1231–47.

64   bullies, who resort to aggression, throwing their weight around, and raw forms of intimidation and dominance, in point of fact, are outcasts and low in the social hierarchy: D. Olweus, "Stability of Aggressive Reaction Patterns in Males: A Review," *Psychological Bulletin* 86 (1979): 852–75.

64   adorn and beautify themselves in an arms race of beauty to attract resource-rich mates: Darwin, *Descent*, chaps. 19–20. For an erudite and illuminating extension of Darwin's ideas, see Geoffrey Miller, *The Mating Mind: How Sexual Choice Shaped the Evolution of Human Nature* (New York: Anchor, 2000). For a playful treatment of the evolutionary biology of

sex, see O. Judson, *Dr. Tatiana's Sex Advice to All Creation* (New York: Owl, 2002).

64 This logic of competing interests extends to parent-offspring relations: Hrdy, *Mother Nature*, chap. 18.

64 This parent-offspring conflict even extends to mother-fetus relations: D. Haig, "Evolutionary Conflicts in Pregnancy and Calcium Metabolism—A Review," *Placenta* 25. *Supplement A. Trophoblast Research* 18 (2004): S10–S15.

65 In an observational study of American families: Judy Dunn and her colleagues have done revealing research on the dynamics of family conflict, and how those conflicts lay a foundation for talks about emotion and morality, and the development of empathy and conflict resolution strategies. J. Dunn and C. Herrera, "Conflict Resolution with Friends, Siblings and Mothers: A Developmental Perspective," *Aggressive Behavior* 23 (1997): 343–57; Dunn and P. Munn, "Becoming a Family Member: Family Conflict and the Development of Social Understanding in the Second Year," *Child Development* 56 (1985): 480–92.

65 This kind of sibling conflict: Frank J. Sulloway, *Born to Rebel: Birth Order, Family Dynamics, and Creative Lives* (New York: Pantheon, 1996).

65 who documented how our primate relatives reconcile after aggressive encounters: Jane Goodall, *The Chimpanzees of Gombe: Patterns of Behavior* (Cambridge, MA: Harvard University Press, 1986). Frans B. M. de Waal, "The Integration of Dominance and Social Bonding in Primates," *Quarterly Review of Biology* 61 (1986): 459–79; de Waal and A. van Roosmalen, "Reconciliation and Consolation among Chimpanzees," *Behavioral Ecology and Sociobiology* 5 (1979): 55–66.

65 the prevailing wisdom: Konrad Lorenz, *On Aggression*, trans. M. Latzke, (London: Methuen, 1967).

66 Recent studies have found that wolves: For an excellent review of the literature on ostracism, see K. Williams, "Ostracism," *Annual Review of Psychology* 58 (2007): 425–52.

66 our survival depends on healthy, stable bonds with others: R. F. Baumeister and M. R. Leary, "The Need to Belong: Desire for Interpersonal Attachments as a Fundamental Human Motivation," *Psychological Bulletin* 117 (1995): 497–529.

66 We are relative prudes compared to these primate relatives: Matt Ridley, *The Red Queen* (New York: HarperCollins, 1993), 213–17.

67 Bonobo females are sexually active for about five years: Frans B. M. de Waal, "Bonobo Sex and Society," *Scientific American* 272 (March 1995): 82–88.

67 This sexual organization had several important implications: Jared Diamond, *Why Is Sex Fun? The Evolution of Human Sexuality* (New York: Basic Books, 1997).

67   sexual monogamy was the most common sexual pattern: Helen Fisher, *Anatomy of Love* (New York: Simon & Schuster, 1993).

68   In *The Evolution of Cooperation:* Robert Axelrod, *The Evolution of Cooperation,* (New York: Basic Books, 1984).

71   Built into the human organism, therefore, must be a set of mechanisms that reverse the cost-benefit analysis of giving: Sober and Wilson, *Unto Others,* chap. 1.

72   Jonathan Haidt has called this state elevation: J. Haidt and D. Keltner, "Appreciation of Beauty and Excellent (Awe, Wonder, Elevation)," in *Character Strengths and Virtues,* ed. Christopher Peterson and Martin E. P. Seligman (New York: Oxford University Press, 2004), 537–51.

EMBARRASSMENT

74   On July 2, 1860, Eadweard Muybridge boarded a stagecoach in San Francisco: Rebecca Solnit, *Rivers and Shadows* (New York: Penguin, 2003).

76   At the time I began my research, the display of embarrassment was thought to be a sign of confusion and thwarted intention: The brilliant sociologist Erving Goffman was fascinated by embarrassment and described it as reflecting a state of confusion. He did, however, suggest that it was a critical signal of an individual's commitment to the social order—an observation that would guide much of the work on embarrassment. Goffman, *Interaction Ritual: Essays on Face-to-Face Behavior* (New York: Doubleday, 1967), 97–112.

76   In a frenzied eighteen months at the University of Pennsylvania: Eadweard Muybridge, *The Human Figure in Motion* (New York: Dover, 1955).

79   That is the orienting function of the startle: S. S. Tomkins, "Affect Theory," in *Approaches to Emotion,* 163–95.

79   the magnitude of the 250-millisecond startle response is a telling indicator of a person's temperament, and in particular of the extent to which the person is anxious: There is now an extensive literature that relies on the magnitude of the startle response, most typically measured in terms of the intensity of the eyeblink, as an index of fear and negative emotion. Just as importantly, a person's positive emotional disposition or current positive emotion tends to attenuate the startle response. P. J. Lang, "The Emotion Probe," *American Psychologist* 50 (1995): 372–85; Lang, M. N. Bradley and B. N. Cuthbert, "Emotion, Attention, and the Startle Reflex," *Psychological Review* 97 (1990): 377–95, and "Emotion, Motivation, and Anxiety: Brain Mechanisms and Psychophysiology," *Biological Psychiatry* 44 (1998): 1248–63.

79   neurotic individuals make for more difficult marriages: Neuroticism is defined by elevated levels of tension, anxiety, worry, and self-doubt. As

important as negative emotions are in certain contexts, their chronic occurrence has proven to be difficult for marriages. N. Bolger and E. A. Schilling, "Personality and the Problems of Everyday Life: The Role of Neuroticism in Exposure and Reactivity to Daily Stressors," *Journal of Personality* 59 (1991): 355–86.

81 My first step was to embarrass people, a task that has given license to a more mischievous side of researchers' imaginations: For reviews of studies of embarrassment, see R. S. Miller, "The Nature and Severity of Self-Reported Embarrassing Circumstances," *Personality and Social Psychology Bulletin* 18 (1992): 190–98; D. Keltner and B. N. Buswell, "Embarrassment: Its Distinct Form and Appeasement Functions," *Psychological Bulletin* 122 (1997): 250–70.

81 after eighteen months of age, they show embarrassment: M. Lewis, M. V. Sullivan, C. Stanger, and M. Weiss, "Self-Development and Self-Conscious Emotions," *Child Development* 60 (1989): 146–56.

81 perhaps the most mortifying experiment: D. Shearn et al., "Facial Coloration and Temperature Responses in Blushing," *Psychophysiology* 27 (1990): 687–93.

82 and one in line with Darwin-inspired analyses of emotional displays as involuntary, truthful signs: Ekman, "Facial Expression and Emotion"; A. J. Fridlund, *Human Facial Expression: An Evolutionary View* (San Diego: Academic Press, 1994). D. Keltner and A. Kring, "Emotion, Social Function, and Psychopathology," *General Psychological Review* 2 (1998): 320–42.

84 Consider the kiss: J. Foer, "The Kiss of Life," *The New York Times*, February 14, 2006.

85 documented by: Eibl-Eibesfeldt, *Human Ethology*.

85 Frans de Waal has devoted thousands of hours to the study of what different primates: de Waal and van Roosmalen, "Reconciliation and Consolation."

86 When I reviewed forty studies of appeasement and reconciliation processes across species: Keltner and Buswell, "Embarrassment: Its Distinct Form."

86 the loss of body control (the prosaic fart or stumble): For one study that has characterized the different causes of embarrassment, see R. S. Miller, "The Nature and Severity of Self-Reported Embarrassing Circumstances," *Personality and Social Psychology Bulletin* 18 (1992): 190–98.

89 I concentrated on young boys prone to violence: D. Keltner, T. Moffitt, and M. Stouthhamer-Loeber, "Facial Expressions of Emotion and Psychopathology in Adolescent Boys," *Journal of Abnormal Psychology* 104 (1995): 644–52.

90 Neuroscientist James Blair has followed up on this work on embarrassment and violence by studying "acquired sociopathy": R. J. R. Blair

and L. Cipolotti, "Impaired Social Response Reversal: A Case of 'Acquired Sociopathy,'" *Brain* 123 (2000): 1122–41.

92   Like J. S., Muybridge had damaged his orbitofrontal cortex, which might be thought of as a command center for the moral sentiments: Edmund T. Rolls, *The Brain and Emotion* (New York: Oxford University Press, 1999).

92   the amygdala, a small, almond-shaped part of the midbrain: Joseph LeDoux, *The Emotional Brain: The Mysterious Underpinnings of Emotional Life* (New York: Simon & Schuster, 1996).

92   It receives information from the cingulate cortex: For a review, see R. J. Davidson, D. Pizzagalli, J. B. Nitschke, and N. H. Kalin, "Parsing the Subcomponents of Emotion and Disorders: Perspective from Affective Neuroscience," in *Handbook of Affective Sciences*, 8–24.

92   Soft, velvety touch to the arm: E. T. Rolls, "The Orbitofrontal Cortex and Reward," *Cerebral Cortex* 10 (2000): 284–94.

93   "He is fitful": J. M. Harlow, "Recovery from the Passage of an Iron Bar through the Head," *History of Psychiatry* 4 (1993): 274–81.

93   In research with Jennifer Beer: J. Beer et al., "The Regulatory Function of Self-Conscious Emotion: Insights from Patients with Orbitofrontal Damage," *Journal of Personality and Social Psychology* 85 (2003): 594–604.

94   They resembled psychopaths: R. J. R. Blair, R. L. Jones, F. Clark, and M. Smith, "The Psychopathic Individual: A Lack of Responsiveness to Distress Cues?" *Psychophysiology* 34, no. 2 (1997): 192–98.

95   "When man is born": Lao Tzu, *Tao Te Ching* (New York: Penguin, 1963), Book II. LXXVI.

SMILE

98   Greek artisans: Agnus Trumble, *A Brief History of the Smile* (New York: Basic Books, 2004), 11–18.

99   What does the smile mean?: M. Frank, P. Ekman, and W. V. Friesen, "Behavioral Markers and Recognizability of the Smile of Enjoyment," *Journal of Personality and Social Psychology* 64 (1993): 83–93; A. J. Fridlund, "Sociality of Solitary Smiling: Potentiation by an Implicit Audience," *Journal of Personality and Social Psychology* 60 (1991): 229–40.

99   If the right kind of smile is synonymous with happiness: D. Keltner et al., "Facial Expression of Emotion," *Handbook of Affective Science*, ed. Richard Davidson, Klaus Scherer, and Hill H. Goldsmith, (London: Oxford University Press, 2003), 415–32.

99   Charles Darwin's analysis of the smile: Darwin, *Expression*, chap. 8.

101   In her careful observations of primates: S. Preuschoft and J. A. R. A. M. Van Hooff, "The Social Function of 'Smile' and 'Laughter:' Varia-

tions across Primate Species and Societies," in *Where Nature Meets Culture: Nonverbal Communication in Social Interaction,* ed. U. Segerstråle and P. Molnár, (Hillsdale, NJ: Erlbaum, 1997), 171–89.

102   I first encountered the deferential smile: D. Keltner et al., "Teasing in Hierarchical and Intimate Relations," *Journal of Personality and Social Psychology* 75 (1998): 1231–47.

104   Research shows that when workers smile in the service industry: For a review of the role of emotion in the workplace, see M. W. Morris and D. Keltner, "How Emotions Work: An Analysis of the Social Functions of Emotional Expression in Negotiations," *Review of Organizational Behavior* 22 (2000): 1–50.

104   workers experience a problematic disconnect: Arlie R. Hochschild, *The Managed Heart: Commercialization of Human Feeling* (Berkeley: University of California Press, 1983).

104   This disconnect has parallels to recent studies by my colleague Ann Kring of schizophrenics: A. M. Kring, S. L. Kerr, A. D. Smith, and J. M. Neale, "Flat Affect in Schizophrenia Does Not Reflect Diminished Subjective Experience of Emotion," *Journal of Abnormal Psychology* 102 (1993): 507–17; Kring and Neale, "Do Schizophrenics Show a Disjunctive Relationship among Expressive, Experiential, and Psychophysiological Components of Emotion?" *Journal of Abnormal Psychology* 105 (1996): 249–57.

105   the empirical literature on the smile yields similarly paradoxical findings: Keltner et al., "Facial Expression of Emotion," in *Handbook of Affective Sciences,* 415–32.

105   The answer is provided by Paul Ekman, and it involves looking away from the lip corners: Frank, Ekman, and Friesen, "Behavioral Markers and Recognizability of the Smile of Enjoyment," *Journal of Personality and Social Psychology* 64 (1993): 83–93.

105   Duchenne smiles differ morphologically: ibid., 83.

106   D smiles tend to be associated with activity in the left anterior portion of the frontal lobes: Neuroscientist Richard Davidson has made the persuasive case that positive emotions tend to activate regions of the brain on the left side of the frontal lobes, because these regions enable the individual to approach rewards. Ekman, Davidson, and Friesen, "The Duchenne Smile: Emotional Expression and Brain Physiology II," *Journal of Personality and Social Psychology* 58 (1996): 342–53.

106   When a ten-month-old is approached by his or her mother: Davidson and N. A. Fox, "Frontal Brain Asymmetry Predicts Infants' Response to Maternal Separation," *Journal of Abnormal Psychology* 98 (1989): 127–31.

107   we interviewed middle-aged adults six months after their deceased spouse had passed away: D. Keltner and G. A. Bonanno, "A Study of

Laughter and Dissociation: The Distinct Correlates of Laughter and Smiling During Bereavement," *Journal of Personality and Social Psychology* 73 (1997): 687–702.

108   In *Emotions Revealed*: Paul Ekman, *Emotions Revealed* (New York: Owl, 2004).

108   In the 1980s developmental psychologists: E. Z. Tronick, "Emotions and Emotional Communications in Infants," *American Psychologist* 44 (1989): 112–19; Tronick, J. Cohn, and E. Shea, "The Transfer of Affect between Mothers and Infants," *Affective Development in Infancy*, ed. T. B. Brazelton and M. W. Yogman (Norwood, NJ: Ablex, 1986); T. Field et al., "Behavior State Matching and Synchrony in Mother-Infant Interactions of Nondepressed Versus Depressed Dyads," *Developmental Psychology* 26 (1990): 7–14; Field et al., "Infants of Depressed Mothers Show 'Depressed' Behavior Even with Nondepressed Adults," *Child Development* 59 (1988): 1569–79.

111   Friends of depressives: Connie Hammen and Ian Gotlib have done superb work documenting the social costs of depression, how it transmits to others and turns relationships into more complex, and at times less rewarding, endeavors. One likely reason is that depressives give off fewer positive emotional cues, in such behaviors as the smile, laugh, and playful touch. Ian H. Gotlib and Constance L. Hammen, *Psychological Aspects of Depression: Toward a Cognitive-Interpersonal Integration* (Chichester: Wiley, 1992).

111   In conversations with individuals who show little positive emotion in the face or voice: For a review of these findings, see Keltner and Kring, "Emotion, Social Function, and Psychopathology."

111   will eagerly cross the surface, risking potential harm, to be in the warm, reassuring midst of their mother's smile: J. F. Sorce, R. N. Emde, Joseph Campos, and M. D. Klinnert, "Maternal Emotional Signaling: Its Effect on the Visual Cliff Behavior of One-Year-Olds," *Developmental Psychiatry* 21 (1985): 195–200; M. D. Klinnert, R. N. Emde, P. Butterfield, J. Campos, "Social Referencing: The Infant's Use of Emotional Signals from a Friendly Adult with Mother Present," *Developmental Psychology* 22 (1986): 427–32.

111   when people emit D smiles when experiencing stress: B. L. Fredrickson and R. W. Levenson, "Positive Emotions Speed Recovery from the Cardiovascular Sequelae of Negative Emotions," *Cognition and Emotion* 12 (1998): 191–220.

111   The definitive work on this topic: For a review of this superb work, see U. Dimberg and A. Öhman, "Behold the Wrath: Psychophysiological Responses to Facial Stimuli," *Motivation and Emotion* 20 (1996): 149–82.

112   suggest that perceiving smiles in others, most likely of the Duchenne variety, triggers the release of the neurotransmitter dopamine:

R. Depue and J. Morrone-Strupinsky, "A Neurobehavioral Model of Affiliative Bonding: Implications for Conceptualizing a Human Trait of Affiliation," *Behavioral and Brain Sciences* 28 (2005): 313–95.

112 As one illustration: C. Senior, "Beauty in the Brain of the Beholder," *Neuron* 38 (2004): 525–28.

114 Undaunted, LeeAnne Harker and I took a week to code the yearbook photos: L. A. Harker and D. Keltner, "Expressions of Positive Emotion in Women's College Yearbook Pictures and their Relationship to Personality and Life Outcomes across Adulthood," *Journal of Personality and Social Psychology* 80 (2001): 112–24.

116 Dozens of scientific studies have found that people who are led to experience brief positive emotions are more creative: Research by Alice Isen and Barbara Fredrickson dispels many myths about the thoughtlessness of positive emotion. Instead, the consistent theme to emerge is that positive emotions make our thought processes more creative and sophisticated. Fredrickson, "What Good Are Positive Emotions?" *Review of General Psychology* 2 (1998): 300–19, and "The Role of Positive Emotions in Positive Psychology: The Broaden-and-Build Theory of Positive Emotions," *American Psychologist* 56 (2001): 218–26; A. M. Isen, K. A. Daubman, and G. P. Nowicki, "Positive Affect Facilitates Creative Problem Solving," *Journal of Personality and Social Psychology* 52 (1987): 1122–31; A. M. Isen, "Positive Affect, Cognitive Processes, and Social Behavior," in *Advances in Experimental Social Psychology*, ed. Leonard Berkowitz (New York: Academic Press, 1987), 203–53.

116 Much has been made of the toxic effects on marriages of negative emotions: For a summary of Gottman and Levenson's research, see Gottman, *Why Marriages Succeed or Fail* (New York: Simon & Schuster, 1993).

116 Here Gottman and colleagues are starting to show: Gottman and R. W. Levenson, "Rebound from Marital Conflict and Divorce Prediction," *Family Processes* 38 (1999): 287–92, and "The Timing of Divorce: Predicting When a Couple Will Divorce over a Fourteen-Year Period," *Journal of Marriage and the Family* 62 (2001): 737–45; Gottman et al., "Predicting Marital Happiness and Stability from Newlywed Interactions," *Journal of Marriage and the Family* 60, no. 1 (1998): 5–22.

117 Physical attractiveness has been shown to have a host of benefits: K. K. Dion, E. Berscheid, and E. Walster, "What Is a Beautiful Good," *Journal of Personality and Social Psychology* 24 (1972): 285–90.

119 For example, Silvan Tomkins: Tomkins, "Affect Theory."

119 For Freud, many pleasurable experiences: Sigmund. Freud, *The Psychopathology of Everyday Life*. The Pelican Freud Library, vol. 5, trans. J. Strachey et al., (Harmondsworth, UK: Penguin, 1975); Anna Freud, *The Ego and the Mechanisms of Defense* (London: Hogarth Press, 1937).

119 Terror management theory, a widely influential theory in social psychology: J. Greenberg et al., "Evidence for Terror Management Theory II: The Effects of Mortality Salience on Reactions to Those Who Threaten or Bolster the Cultural Worldview," *Journal of Personality and Social Psychology* 58 (1990): 308–18.

120 It is assumed in the study of parent-child attachment: M. Mikulincer and P. R. Shaver, "The Attachment Behavioral System in Adulthood: Activation, Psychodynamics, and Interpersonal Processes," in *Advances in Experimental Social Psychology*, vol. 35, ed. Mark P. Zanna (New York: Academic Press, 2003), 53–152.

120 The Woody Allen hypothesis has deep roots in Judeo-Christian thought about original sin and the fall from grace: One of the best books I've read on happiness is historian Darrin McMahin's broad survey of how the concept and practice of happiness have changed over 2,500 years of Western culture. McMahin, *Happiness: A History* (New York: Grove Press, 2006).

121 This is a standard evolutionary principle: R. J. Andrew, "The Origin and Evolution of the Calls and Facial Expressions of the Primates," *Behavior* 20 (1963): 1–109. P. Rozin, "Towards a Psychology of Food and Eating: From Motivation to Module to Model to Marker, Morality, Meaning, and Metaphor," *Current Directions in Psychological Science* 5 (1996): 18–24.

121 downplay any sudden abundance in resources through modesty and generosity: Christopher Boehm, *Hierarchy in the Forest*.

LAUGHTER

123 Jared Diamond argues: Jared Diamond, *Guns, Germs, and Steel: The Fate of Human Societies* (New York: W. W. Norton, 1997).

124 It is a point that evolutionists Matthew Gervais and David Sloan Wilson: Gervais and Wilson, "The Evolution and Functions of Laughter: A Synthetic Approach," *Review of Quarterly Biology* 80 (2006): 395–430.

124 What separates mammals from reptiles are the raw materials of laughter: MacLean was one of the first to make this point in explaining how mammals' brains differ from those of reptiles. Paul D. MacLean, *The Triune Brain in Evolution* (New York: Plenum, 1990).

124 Yet the laughter of chimps and apes is more tightly linked to inhalation and exhalation patterns: Robert Provine has written a wonderful book on laughter. Provine, *Laughter: A Scientific Investigation* (New York: Viking, 2000).

125 Boyle's descriptions: T. C. Boyle, *Drop City* (New York: Penguin, 2003), various pages.

127 Estimates indicate that: Provine, *Laughter*.

127 Laughter is contagious: Robert Provine, "Contagious Laughter: Laughter Is a Sufficient Stimulus for Laughs and Smiles," *Bulletin of the Psychonomic Society* 30 (1992): 1–4. Provine also details some wonderful episodes of contagious laughter in *Laughter*.

128 laughter is intertwined with our breathing: Provine was the first to make this point, in *Laughter*.

130 Measures include speech rate, pitch, loudness: Klaus Scherer is to the voice in emotion what Ekman is to the face. He was the first to chart systematically how the different emotions will be expressed in difficult acoustic qualities like pitch, amplitude, and variation. K. R. Scherer, "Vocal Affect Expression: A Review and a Model for Future Research," *Psychological Bulletin* 99 (1986): 143–65.

130 Bachorowski was the first to put laughs through this complex form of acoustic analysis: Jo-Anne Bachorowski has done for the laugh what Ekman did for the smile—provide an objective, anatomically based rationale for explaining some of the varieties of the laugh. Bachorowski and M. J. Owren, "Not All Laughs Are Alike: Voiced by Not Voiced Laughter Readily Elicits Positive Affect," *Psychological Science* 12 (2001): 252–57.

130 In his remarkable meditation on laughter: Milan Kundera, *The Book of Laughter and Forgetting* (New York: Penguin, 1981).

131 The laughs of friends, as opposed to those of strangers: Smoski and Bachorowski, "Antiphonal Laughter between Friends and Strangers," *Cognition and Emotion* 17 (2003): 327–40.

131 Here a remarkable discovery: Bachorowski, Smoski, and Owren, "The Acoustic Features of Human Laughter," *Journal of Acoustic Society of America* 110 (2001): 1581–97.

131 laughter preceded language in human evolution: For analyses of the evolution of laughter, see Provine, *Laughter*; Gervais and Wilson, "The Evolution and Functions of Laughter."

131 Recent neuroscientific data on laughter: B. Wild, F. A. Rodden, W. Grodd, and W. Ruch, "Neural Correlates of Laughter and Humor," *Brain* 126, no. 10 (2002): 2121–38.

132 tension and ambiguity: for an excellent treatment of humor and laughter, see Michael L. Apte, *Humor and Laughter: An Anthropological Approach* (Ithaca: Cornell University Press, 1985).

132 Provine turned his astute ear to the laughter that occurs in the real world: Robert Provine, "Laughter Punctuates Speech: Linguistic, Social, and Gender Contexts of Laughter," *Ethology* 95 (1993): 291–98.

134 the answer is cooperation: Owren and Bachorowski, "The Evolution of Emotional Expression: A 'Selfish-Gene' Account of Smiling and Laughter in Early Hominids and Humans, in *Emotions: Current Issues and Future Directions*, ed. Tracy J. Mayne and George A. Bonanno (New York: Guilford Press, 2001), 151–91.

134 The first is contagion: Robert Provine details one of my favorite examples of contagious laughter. In a grammar school, girls in one class started laughing, and this outbreak spread throughout the school. Girls laughed for days until the school had to be closed. Notwithstanding the difficulties this contagious laughter produced for the conduct of class, these girls, Bachorowski and Owren argue, are bonding through the contagious delights of laughter.

134 when we hear others laugh, mirror neurons represent that expressive behavior: N. Osaka et al., "An Emotion-Based Facial Expression Word Activates Laughter Module in the Human Brain: A Functional Magnetic Resonance Imaging Study," *Neuroscience Letters* 340, no. 2 (2003): 127–30.

135 In my own research with executives: For the past fifteen years, I have videotaped executives from around the world conducting negotiations with other executives, and then coded those videotapes for different emotional displays. As reliable as the handshake to begin the negotiation and dramatic displays of anger and contempt when the tension mounts is the occurrence of laughter in the initial stages. This laughter paves the way for increased trust and more integrative bargaining, where the two parties more effectively understand and act upon their respective interests. M. W. Morris and D. Keltner "How Emotions Work: An Analysis of the Social Functions of Emotional Expression in Negotiations," *Review of Organizational Behavior* 22 (2000): 1–50.

135 Workplace studies find that coworkers often laugh when negotiating potential conflicts: R. L. Coser, "Laughter among Colleagues," *Human Relations* 12 (1960): 171–82.

135 Strangers who laugh while flirting: K. Grammer, "Strangers Meet: Laughter and Nonverbal Signs of Interest in Opposite-Sex Encounters," *Journal of Nonverbal Behavior* 14 (1990): 209–36.

135 Friends whose laughs join in antiphonal form: Smoski and Bachorowski, "Antiphonal Laughter."

135 For couples who divorced on average 13.9 years after they were married: While much has been made of the toxic effects of negative emotions such as contempt and criticism in marriage, Gottman and Levenson and others have begun to look at the benefits of positive emotions, such as mirth and laughter. In their writing about laughter, Gottman and Levenson suggest that problematic discussions in intimate life are like negative affect cascades—feelings of anger and resentment rise and build upon one another. Couples who can exit from these cascades fare much better, and one manner of exiting is laughter. Gottman and Levenson, "Timing of Divorce."

136 Hobbes: *Leviathan*, ed. Richard Tuck (Cambridge: Cambridge University Press, 1991), 43.

136 It is for these reasons that Steve Pinker called: Steven Pinker, *The Language Instinct* (New York: HarperCollins, 1994).

137 In his analysis of the development of pretense: A. M. Leslie, "Pretense and Representation: The Origins of 'Theory of Mind,'" *Psychological Review* 94, no. 4 (1987): 412–26.

138 Linguist Paul Drew carefully analyzed the unfolding of family teasing: P. Drew, "Po-Faced Receipts of Teases," *Linguistics* 25 (1987): 219–53.

140 For the past fifteen years: G. A. Bonanno and S. Kaltman, "Toward an Integrative Perspective on Bereavement," *Psychological Bulletin* 125 (1999): 760–76.

140 To test this thesis, George and I undertook a study: G. A. Bonanno and D. Keltner, "Facial Expressions of Emotion and the Course of Conjugal Bereavement," *Journal of Abnormal Psychology* 106 (1997): 126–37.

143 We coded participants' references to several existential themes related to bereavement: G. A. Bonanno and D. Keltner, "The Coherence of Emotion Systems: Comparing 'On-Line' Measures of Appraisal and Facial Expressions, and Self-Report," *Cognition and Emotion* 18 (2004): 431–44.

TEASE

146 Not so, reason: The Zahavis make a wonderful case for the role of provocation in animal world. They suggest that, as with humans, nonhuman species often need to assess each other's commitments, and they do so through provocation. This analysis is very much in keeping with Robert Frank's analysis of the importance of emotion in motivating commitments, and was central to how members of my lab thought about the functions of teasing in human social life. A. Zahavi and A. Zahavi, *The Handicap Principle* (New York: Oxford University Press, 1997).

147 Chimpanzees dangle their tails: O. Adang, "Harassment of Mature Female Chimpanzees by Young Males in the Mahale Mountains," *International Journal of Primatology* 24 (2003): 503–14.

147 Adults will play hide the face: Bambi B. Schieffelin, *The Give and Take of Everyday Life: Language Socialization of Kaluli Children* (Cambridge: Cambridge University Press, 1990).

147 Teenage girls and boys resort to hostile nicknames and outlandishly gendered imitations: Barrie Thorne, *Gender Play: Girls and Boys in School* (New Brunswick, NJ: Rutgers University Press, 1993); D. Eder, "The Role of Teasing in Adolescent Peer Group Culture," *Sociological Studies of Child Development* 4 (1991): 181–97, and "'Go Get Ya a French!': Romantic and Sexual Teasing Among Adolescent Girls," *Gender and Conversational Interaction: Oxford Studies in Sociolinguistics*, ed. Deborah Tannen (New York: Oxford University Press, 1993), 17–31.

148 Teasing has long occupied a problematic place in Western culture:

For such an analysis, see Keltner et al., "Just Teasing: A Conceptual Analysis and Empirical Review," *Psychological Bulletin* 127 (2001): 229–48. The same ambivalence is evident in Western culture's stances toward two close relatives of teasing, satire and irony. For excellent cultural, historical analyses of satire and irony and their shifting prominence in Western culture, see Duncan Griffin, *Satire: A Critical Reintroduction* (Lexington, KY: University of Kentucky Press, 1994), and Linda Hutcheon, *Irony's Edge* (London: Routledge, 1995).

148   the most sterling of reputations: Hutcheon, *Irony's Edge*, 7.

148   issued a clarion call: Jedidiah Purdy, *For Common Things: Irony, Trust, and Commitment in America Today* (New York: Random House, 2000).

148   On January, 19, 1449, the Scots passed: For an outstanding history of the fool, and the surprising power that fools have enjoyed until recently, see Bernice K. Otto, *Fools Are Everywhere* (Chicago: University of Chicago Press, 2001).

149   The prominence of the jester and fool: For an excellent anthropological history of the fool, see Apte, *Humor and Laughter.*

150   The scientific study of teasing was hampered by poorly specified definitions: Keltner et al., "Just Teasing."

150   In terms more felicitous to scientific inquiry: Ibid.

151   philosopher Paul Grice outlined four principles of communication: H. P. Grice, *Logic and Conversation* (New York: Academic Press, 1975).

152   The relevance of Grice's maxims to teasing, ironically enough: Penelope Brown and Stephen C. Levinson, *Politeness: Some Universals in Language Usage* (Cambridge: Cambridge University Press, 1987).

153   that act is fraught with potential conflict: Brown and Levinson drew heavily upon the brilliant insights of Erving Goffman. Goffman offered a strategic or dramaturgic account of human social life, arguing that many of the rituals and practices of social life are organized to protect our sense of social esteem, or what he called "face." Goffman is required reading to understand the universal tendencies toward politeness, modesty, and deference. He reveals the profound degree of cooperation in our public life. See *The Presentation of Self in Everyday Life* (Garden City: Doubleday, 1959), *Behavior in Public Places* (New York: Free Press, 1966), and *Interaction Ritual: Essays on Face-to-Face Behavior* (New York: Doubleday, 1967).

153   the mother referred to a young son as "horse mouth": E Ochs, introduction to *Language Socialization across Cultures: Studies in the Social and Cultural Foundations of Language,* vol. 3, ed. B. B. Schieffelin and E. Ochs (New York: Cambridge University Press, 1986), 1–16.

154   Exaggeration is core to understanding "playing the dozens": R. D. Abrahams, "Playing the Dozens," *Journal of American Folklore* 75 (1962): 209–20.

155 "here's your dog food": C. A. Straehle, " 'Samuel?' 'Yes, dear?' Teasing and Conversational Rapport," *Gender and Conversational Interaction: Oxford Studies in Sociolinguistics* (New York: Oxford University Press, 1993), 210–30.

155 When we tease, linguist Herb Clark observes, we frame the interaction as one that occurs in a playful, nonserious realm of social exchange: Herbert H. Clark, *Using Language* (Cambridge: Cambridge University Press, 1996). Clark has been exploring how language offers the capacity to represent different layers of social reality, and he suggests that in teasing we take on different identities.

156 The philosopher Bertrand Russell argued: Bertrand Russell, *Power: A New Social Analysis* (London: Allen and Unwin, 1938), 10.

157 Male fig wasps: Krebs and Davies, *An Introduction to Behavioural Ecology*, 157.

157 Given the enormous costs of negotiating rank, many species have shifted to ritualized battles: Ibid., chap. 7.

157 frogs and toads use the depth of their croaks to negotiate rank: ibid.

158 boys who were rising to the top of the hierarchy: R. C. Savin-Williams, "Dominance in a Human Adolescent Group," *Animal Behavior* 25 (1977): 400–406.

159 Our nickname paradigm: Keltner et al., "Teasing in Hierarchical and Intimate Relations."

160 The great satirist Rabelais described nicknames: F. Rabelais, *Garantua and Pantagruel*, trans. J. Cohen (Baltimore: Penguin/Everyman's Library, 1955).

164 Monica Moore surreptitiously observed: M. M. Moore, "Nonverbal Courtship Patterns in Women: Context and Consequences," *Ethology and Sociobiology* 6 (1985): 237–47.

165 partners with a richer vocabulary of teasing insults are happier: R. A. Bell, N. L. Buerkel-Rothfuss, and K. E. Gore, " 'Did You Bring the Yarmulke for the Cabbage Patch Kid?' The Idiomatic Communication of Young Lovers," *Human Communication Research* 14, no. 1 (1985): 47–67; L. A. Baxter, "An Investigation of Compliance-Gaining as Politeness," *Human Communication Research* 10 (1984): 427–56; L. A. Baxter, "Forms and Functions of Intimate Play in Personal Relationships," *Human Communication Research* 18 (1984): 336–63.

165 couples who had been together for several years tease each other: Keltner et al., "Teasing in Hierarchical and Intimate Relations."

166 thanks to research by Craig Anderson and Brad Bushman: C. A. Anderson and B. J. Bushman, "Effects of Violent Video Games on Aggressive Behavior, Aggressive Cognition, Aggressive Affect, Physiological Arousal, and Prosocial Behavior: A Meta-Analytic Review of the Scientific Literature," *Psychological Science* 12 (2001): 353–59.

167   It emerges early: V. Reddy, "Playing with Others' Expectations: Teasing and Mucking About in the First Year," in *Natural Theories of Mind: Evolution, Development, and Simulation of Everyday Mindreading*, ed. A. Whiten (Oxford: Blackwell, 1991), 143–58.

167   The teasing of children with obesity problems: J. K. Thompson, J. Cattarin, B. Fowler, and E. Fisher, "The Perception of Teasing Scale (POTS): A Revision and Extension of the Physical Appearance Related Teasing Scale (PARTS)," *Journal of Personality Assessment* 65 (1995): 146–57; Thompson et al., "Development and Validation of the Physical Appearance Related Teasing Scale," *Journal of Personality Assessment* 56 (1991): 513–21.

167   The literature on bullies bears this out: D. Olweus, *Aggression in Schools: Bullies and Whipping Boys* (New York: Wiley and Sons, 1978); Olweus, *Bullying at School: What We Know and What We Can Do* (Cambridge, MA: Blackwell, 1993).

168   elevated love, amusement, and mirth: Keltner et al., "Teasing in Hierarchical and Intimate Relations."

168   consistent with the tendency for low power to trigger a threat system: D. Keltner, D. H. Gruenfeld, and C. Anderson, "Power, Approach, and Inhibition," *Psychological Review* 110, no. 2 (2003): 265–84.

168   teasing in romantic bonds defined by power asymmetries: ibid.

168   they add irony and sarcasm to their social repertoire: E. Winner, *The Point of Words: Children's Understanding of Metaphor and Irony* (Cambridge, MA: Harvard University Press, 1988), Winner and S. Leekam, "Distinguishing Irony from Deception: Understanding the Speaker's Second-Order Intention," *British Journal of Developmental Psychology* 9 (1991): 257–70.

168   a precipitous twofold drop in the reported incidences of bullying: P. K. Smith and P. Brain, "Bullying in Schools: Lessons from Two Decades of Research," *Aggressive Behavior* 26, no. 1 (2000): 1–9.

168   we created an opportunity for boys at two different developmental stages to taunt one another at a basketball camp: M. A. Logli et al., "Teasing, Taunting, and Gossip," unpublished manuscript.

170   There are still many mysteries to Asperger's Syndrome: Marian Sigman and Lisa M. Capps, *Children with Autism* (Cambridge, MA: Harvard University Press, 1997); U. Frith, F. Happe, and F. Siddons, "Autism and Theory of Mind in Everyday Life," *Social Development* 3, no. 2 (1994): 108–24.

170   as revealed in the brilliant essay by music critic Tim Page: T. Page, "Parallel Play: A Lifetime of Restless Isolation Explained," *The New Yorker*, August 20, 2007.

171   And teasing: E. A. Heerey et al., "Understanding Teasing: Lessons

from Children with Autism," *Journal of Child Abnormal Psychology* 33 (2005): 55–68.

TOUCH

173   For the past fifteen years: mind and life dialogues, www.mindand life.org.

173   the Dalai Lama has been engaging: His Holiness the Dalai Lama, "Understanding Our Fundamental Nature," in *Visions of Compassion: Western Scientists and Tibetan Buddhists,* ed. R. Davidson and A. Harrington (New York: Oxford University Press, 2001), 66–80.

177   an answer is found in the contagious goodness hypothesis: I summarize these ideas in greater detail elsewhere. D. Keltner, "The Compassionate Instinct," *Greater Good* 1 (2004): 6–9. The ideas I summarize as part of the viral goodness hypothesis found inspiration in several sources. Robert Axelrod, *The Evolution of Cooperation* (New York: Basic Books, 1971); R. L. Trivers, "The Evolution of Reciprocal Altruism," *Quarterly Review of Biology* 46 (1984): 35–57; Frank, *Passions Within Reason*; Sober and Wilson, *Unto Others*.

178   Desmond Morris's famous phrasing: Desmond Morris, *The Naked Ape: A Zoologist's Study of the Human Animal* (New York: Dell, 1967).

178   As Nina Jablonski has argued: Jablonski, *Skin*, 39–42.

179   several functions essential to human survival: For an outstanding summary of the function of the skin, see ibid.

180   we learned to signal different objects and states with what are known as emblems: For one of the first taxonomies of expressive behavior, including gestures, see Ekman and Friesen, "The Repertoire of Nonverbal Behavior: Categories, Origins, Usage and Coding," *Semiotica* 1 (1969): 49–98.

181   The progenitor of this view: Rolls, *The Brain and Emotion*.

181   in one study participants received a fifteen-minute Swedish massage: R. A. Turner et al., "Preliminary Research on Plasma Oxytocin in Normal Cycling Women: Investigating Emotion and Interpersonal Distress," *Psychiatry: Interpersonal and Biological Processes* 62 (1999): 97–113.

181   Other studies have found that massage: For a terrific summary of all facets of touch, see Tiffany Field, *Touch* (Cambridge, MA: MIT Press, 2001).

181   Recent studies have found that rat mothers: D. Francis and M. J. Meaney, "Maternal Care and the Development of Stress Responses," *Development* 9 (1999): 128–34.

182   Dopamine is a neurotransmitter that is involved in the pursuit of rewards: A. G. Phillips et al., "Neurobiological Correlates of Positive Emotional States: Dopamine, Anticipation, and Reward," in *International Review of Studies on Emotion*, vol. 2, ed. K. T. Strongman (New York: John Wiley, 1992), 31–50.

182   the benefits of touching are not limited to rat pups: For a comprehensive review of the dozens of studies of touch therapies, and the benefits they bring to premature babies, people suffering from depression, and ailments during aging, see Field, *Touch*.

182   when teachers are randomly assigned to touch some of their students and not others: D. C. Aguilera, "Relationship between Physical Contact and Verbal Interaction between Nurses and Patients, *Journal of Psychiatric Nursing and Mental Health Services* 5 (1967): 5–21; J. D. Fisher, M. Rytting, and R. Heslin, "Hands Touching Hands: Affective and Evaluative Effects of an Interpersonal Touch," *Sociometery* 39 (1976): 416–21. For a review of the effects of touch, see M. J. Hertenstein et al., "The Communicative Functions of Touch in Humans, Nonhuman Primates, and Rats: A Review and Synthesis of the Empirical Research," *Genetic, Social, and General Psychology Monographs* 132 (2006): 5–94.

182   In her excellent book: Field, *Touch*.

183   mortality rates for infants: Ibid.

183   In a more systematic comparison: R. Spitz, "Hospitalism," *Psychoanalytic Study of the Child* 1 (1945): 53–74.

183   Tiffany Field has found that massages given to premature babies lead, on average, to a 47 percent increase in weight gain: Field reviews several studies, including those from her lab, which replicate this important finding. Field, *Touch*.

183   The infants who were touched during the procedure: L. Gray et al., "Skin-to-Skin Contact Is Analgesic in Healthy New-Borns," *Pediatrics* 105 (2000): 14–20.

183   Touch alters not only our stress-related physiology: Francis and Meaney, "Maternal Care and the Development of Stress Responses."

185   Jim Coan and Richie Davidson had participants wait for a painful burst of white noise: J. A. Coan, H. S. Schaefer, and R. J. Davidson, "Lending a Hand: Social Regulation of the Neural Response to Threat," *Psychological Science* 17 (2006): 1032–39.

187   Frans de Waal, who has studied the role of touch in the patterns of food exchange in chimpanzees: de Waal, *Good Natured*, 136–44.

187   participants were asked to sign a petition in support of a particular issue of local importance: F. N. Willis and H. K. Hamm, "The Use of Interpersonal Touch in Securing Compliance," *Journal of Nonverbal Behavior* 5, no. 1 (1980): 49–55.

187   In a recent study, Robert Kurzban: Kurzban, "The Social Psy-

chophysics of Cooperation: Nonverbal Communication in a Public Goods Game," *Journal of Nonverbal Behavior* 25 (2001): 241–59.

188 catalogued greeting rituals with surreptitious photography in remote cultures: Eibl-Eibesfeldt, *Human Ethology.*

189 when people feel sympathy and are inclined to help others in need, they show a concerned eyebrow and pressed lip: N. Eisenberg et al., "Relation of Sympathy and Distress to Prosocial Behavior."

189 When I presented images of this display: J. Haidt and D. Keltner, "Culture and Facial Expression: Open Ended Methods Find More Faces and a Gradient of Universality," *Cognition and Emotion* 13 (1999): 225–66.

189 I turned to the next best studied modality of emotional communication: E. Simon-Thomas, D. Sauter, and Dacher Keltner, "Vocal Bursts Communicate Distinct Positive Emotions," unpublished manuscript.

190 "Touch is both the alpha and omega": James, *The Principles of Psychology*, vol. 2, 551.

190 So Matt and I designed an experiment: Matthew Hertenstein et al., "Touch Communicates Distinct Emotions," *Emotion* 6 (2006): 528–33.

194 This led Robin Dunbar: Robin I. M. Dunbar, *Grooming, Gossip and the Evolution of Language* (London: Faber and Faber, 1996).

195 We live in a touch-deprived culture: Ashley Montagu, "Animadversions on the Development of a Theory of Touch," in *Touch in Early Development*, ed. Tiffany M. Field (Mahwah, NJ: Erlbaum, 1995) 1–10.

195 "There is a sensible way": J. Watson, *Psychological Care of Infant and Child* (New York: W. W. Norton, 1928), 9–10.

195 In a recent observational study: S. M. Jourard, "An Exploratory Study of Body Accessibility," *British Journal of Social and Clinical Psychology* 5 (1966): 221–31.

196 Compared to infants carried in harder: E. Anisfeld et al., "Does Infant Carrying Promote Attachment? An Experimental Study of the Increased Physical Contact on the Development of Attachment," *Child Development* 61 (1990): 1617–27.

LOVE

200 so sharply summarized in: Ridley, *The Red Queen*, chaps. 6, 7; Cronin, *The Ant and the Peacock*, chaps. 7, 8.

201 universality of serial monogamy: David M. Buss, *The Evolution of Desire: Strategies of Human Mating* (New York: Basic Books, 1994).

201 human males actively contribute to the raising of the offspring: Hrdy, *Mother Nature*; 205–17, Konner, *The Tangled Wing*, 263, 266.

203 spur the scientific study of parent-child love: John Bowlby, *Attachment and Loss*, vols. 1 and 2 (London: Hogarth Press, 1978); Bowlby, *The Making and Breaking of Affectional Bonds* (London: Tavistock, 1979);

Bowlby, *A Secure Base: Clinical Applications of Attachment Theory* (London: Routledge, 1988).

203 she documented familial universals: Mary D. S. Ainsworth, *Infancy in Uganda: Infant Care and the Growth of Love* (Baltimore: Johns Hopkins University Press, 1967).

204 rhesus monkeys raised in isolation: H. F. Harlow, "Love in Infant Monkeys," *Scientific American* 200 (1959): 68–74. H. F. Harlow and M. K. Harlow, "Social Deprivation in Monkeys," *Scientific American* 207 (1962): 136–46.

204 These early attachment experiences, dozens of human studies show, lay the foundation of the capacity to connect: Mario Mikulincer and Philip Shaver have been the leading scientists in extending Bowlby's theorizing to adult human relationships. Mikulincer and Shaver, "The Attachment Behavioral System in Adulthood: Activation, Psychodynamics, and Interpersonal Processes," *Advances in Experimental Social Psychology*, vol. 35, ed. Mark P. Zanna (New York: Academic Press, 2003), 53–152, and *Attachment Patterns in Adulthood: Structure, Dynamics, and Change* (New York: Guilford Press, in press).

204 People who report a sense of secure attachment perceive their partners to be a steady source of support and love: Nancy Collins was one of the first to do the difficult empirical work of applying Bowlby's claims about attachment to the processes of intimate romantic relations. N. L. Collins, "Working Models of Attachment: Implications for Explanation, Emotion, and Behavior," *Journal of Personality and Social Psychology* 71 (1996): 810–32; Collins and B. C. Feeney, "A Safe Haven: An Attachment Theory Perspective on Support Seeking and Caregiving in Intimate Relationships," *Journal of Personality and Social Psychology* 78 (2000): 1053–73.

204 And as life progresses: For an excellent summary of the life-courses of people with different attachment styles, see Mikulincer and Shaver, "The Attachment Behavioral System in Adulthood."

204 A quick study of a morning in such a house: ibid.

205 When Chris Fraley and Phil Shaver surreptitiously observed romantic partners as they said good-bye in airports: R. C. Fraley and Shaver, "Airport Separations: A Naturalistic Study of Adult Attachment Dynamics in Separating Couples," *Journal of Personality and Social Psychology* 75 (1998): 1198–1212.

205 Anxiously attached individuals are more likely to interpret life events: Mikulincer and Shaver, "The Attachment Behavioral System in Adulthood."

207 And bonobos wage: de Waal, "Bonobo Sex and Society."

207 is our love of meat: Ridley, *The Red Queen*, 190.

207   win in the game of sperm competition with other males: ibid. 213–16.

208   the same was happening in human evolution: Jared Diamond, *Why Is Sex Fun?*, 72–93.

208   pole dancers earn bigger tips: G. Miller, J. M. Tybur, and B. D. Jordan, "Ovulatory Cycle Effects on Tip Earnings by Lap Dancers: Economic Evidence for Human Estrus," *Evolution and Human Behavior* 27 (2007): 375–81.

208   The specific language of desire: For superb descriptions of the language of flirtation, see D. B. Givens, *Love Signals: How to Attract a Mate* (New York: Crown, 1983); T. Perper, *Sex Signals: The Biology of Love* (Philadelphia: ISI Press, 1985); Eibl-Eibesfeldt, *Love and Hate: The Natural History of Behavior Patterns*, trans. G. Stracham (New York: Schocken, 1974), chap. 3.

209   These brief signals honor time-honored principles in the game of sexual selection: For an excellent account of the evolution of beauty, see Nancy Etcoff, *Survival of the Prettiest: The Science of Beauty* (New York: Doubleday, 1999). For a more general treatment of how sexual selection processes have led to the evolution of a broader array of behaviors, from humor to music, that bind women and men together, see Miller, *The Mating Mind*.

209   offset by a man of means: David Buss has done controversial and groundbreaking work on how the preferences for beauty and resources shape the mate selection preferences of men and women, respectively. Buss, *The Evolution of Desire*.

210   This kind of behavioral synchrony creates a sense of similarity, trust, and merging of self and other: Elaine Hatfield, John T. Cacioppo, and Ronald L. Rapson, *Emotional Contagion* (Cambridge: Cambridge University Press, 1994).

210   courtship behaviors stimulate the biology of reproduction: Ridley, *The Red Queen*, 81–82.

210   A metaphorical switch in the mind is turned on: These metaphors of love have been documented by George Lakoff and his colleagues in their work on metaphor. The nature of these metaphors closely tracks our experience of romantic love—the voice or rationality is diminished, we feel out of our minds. Clearly the mind is aiding in promoting the kind of devotion required of long-term intimate bonds. Lakoff and Johnson, *Metaphors We Live By;* Lakoff, *Women, Fire and Dangerous Things;* Kövesces, *Metaphor.*

210   And alongside desire, our research finds, they will feel a deep sense of anxiety: When we studied young couples' playful, loving exchanges with each other, we found that they reported high levels of desire (no surprise

there), but those feelings of passion closely corresponded to feelings of anxiety. G. C. Gonzaga et al., "Love and the Commitment Problem in Romantic Relations and Friendship," *Journal of Personality and Social Psychology* 81 (2001): 247–62.

210   the male caricatured all too readily in scientific research: David Buss has done numerous studies that reveal that young men are all too ready to have casual sex. Buss, *The Evolution of Desire*. See also Bruce Ellis, "The Evolution of Sexual Attraction: Evaluative Mechanisms in Women," in *The Adapted Mind*.

210   75 percent of college males: R. D. Clark and E. Hatfield, "Gender Differences in Receptivity to Sexual Offers," *Journal of Psychology and Human Sexuality* 2 (1989): 39–55.

211   they'll often go down in flames of hatred and litigation: David G. Myers, *The American Paradox* (New Haven, CT: Yale University Press, 2000), chap. 3.

211   estimates of adultery: "Adultery Survey Finds 'I do' Means 'I do,'" *New York Times*, October 19, 1993. Although this reference is a bit old, it cites some of the best survey data on the question of rates of adultery.

211   recent studies of abstinence programs provided to middle- and high-school students: www.siecus.org/media/press/press0141.html.

212   In the depths of romantic love, we idealize our partners: Murray and Holmes, "Seeing Virtues in Faults: Negativity and the Transformation of Interpersonal Narratives in Close Relationships," *Journal of Personality and Social Psychology* 65 (1993): 707–23; "A Leap of Faith? Positive Illusions in Romantic Relationships," *Personality and Social Psychology Bulletin* 23 (1997): 586–604; "The (Mental) Ties That Bind: Cognitive Structures That Predict Relationship Resilience," *Journal of Personality and Social Psychology* 77 (1999): 1228–44. Murray et al., "What the Motivated Mind Sees: Comparing Friends' Perspectives to Married Partners' Views of Each Other," *Journal of Experimental Social Psychology* 36 (2000): 600–620; Murray, Holmes, and D. W. Griffin, "The Benefits of Positive Illusions: Idealization and the Construction of Satisfaction in Close Relationships," *Journal of Personality and Social Psychology* 70 (1996): 79–98.

213   More dramatically, romantic love deactivates threat detection: H. E. Fisher, A. Aron, and L. L. Brown, "Romantic Love: A Mammalian Brain System for Mate Choice," *Philosophical Transactions of the Royal Society, Biological Sciences*, 361, no. 1476 (2006): 2173–86.

213   We can pin our hopes on oxytocin: For summaries of what is known about oxytocin, see K. U. Morberg, *The Oxytocin Factor* (Cambridge, MA: De Capo, 2003); S. E. Taylor et al., "Biobehavioral Responses to Stress in Females: Tend-and-Befriend, Not Fight-or-Flight," *Psychological Review* 107 (2000): 411–29; Jaak Panksepp, *Affective Neuroscience*:

The Foundations of Human and Animal Emotions (New York: Oxford University Press, 1998).

213   This remarkable discovery emerged: C. S. Carter, "Neuroendocrine Perspectives on Social Attachment and Love," *Psychoneuroendocrinology* 23, no. 8 (1998): 779–818; J. R. Williams et al., "Oxytocin Administered Centrally Facilitates Formation of a Partner Preference in Female Prairie Voles (Microtus Ochrogaster)," *Journal of Neuroendocrinology* 6 (1998): 247–50.

213   oxytocin increases after sexual behavior: M. S. H. Carmichael et al., "Plasma Oxytocin Increases in the Human Sexual Response," *Journal of Clinical Endocrinology and Metabolism* 64, no. 1 (1987): 27–31.

213   injections of oxytocin increase social contact and pro-social behavior: D. M. Witt, C. Carter, and C. Walton, "Central and Peripheral Effects of Oxytocin Administration," *Physiology and Behavior* 37 (1990): 63–9; Witt, J. T. Winslow, and Thomas Insel, "Enhanced Social Interaction in Rats Following Chronic, Centrally Infused Oxytocin," *Pharmacology, Biochemistry, and Behavior* 43 (1992): 855–61.

213   decrease threatening facial displays: S. D. Holman, and R. W. Goy, "Experiential and Hormonal Correlates of Care-giving in Rhesus Macaques," in *Motherhood in Human and Nonhuman Primates: Biosocial Determinants*, ed. C. R. Pryce and R. D. Martin (Basel: Karger, 1995), 87–93.

214   Little domestic chicks: J. Panksepp, E. Nelson, and M. Bekkedal, "Bain Systems for the Mediation of Social Separation-Distress and Social-Reward: Evolutionary Antecedents and Neuropeptide Intermediaries," in *The Integrative Neurobiology of Affiliation*, ed. C. S. Carter, I. I. Lederhendler, and B. Kirpatrick (New York: New York Academy of Sciences, 1997), 78–100.

214   the physiological underpinnings of love, devotion, and trust: For an accessible review of the literature on oxytocin, see K. U. Morberg, *The Oxytocin Factor* (Cambridge, MA: De Capo, 2003), 105–32.

214   In studies of lactating women: E. B. Keverne, "Psychopharmacology of Maternal Behavior," *Journal of Psychopharmacology* 10 (1996): 16–22; C. S. Carter and M. Altemus, "Integrative Functions of Lactational Hormones in Social Behavior and Stress Management," in *Annals of The New York Academy of Sciences*, vol. 807, ed. Carter and Lederhendler (New York: New York Academy of Sciences, 1997), 164–74.

214   Prepartum mothers who show higher baseline levels of oxytocin later showed increased attachment-related behavior: R. Feldman et al., "Evidence for a Neuroendocrinological Foundation of Human Affiliation: Plasma Oxytocin Levels Across Pregnancy and the Postpartum Period Predict Mother-Infant Bonding," *Psychological Science* 18, no. 11 (2007): 965–70.

215 Gian Gonzaga and I undertook a Darwinian study of sexual desire and romantic love: Gonzaga et al., "Love and the Commitment Problem in Romantic Relations and Friendship."

217 We next turned to a query of our chemical quarry, oxytocin: G. C. Gonzaga et al., "Romantic Love and Sexual Desire in Close Bonds," *Emotion* 6 (2006): 163–79.

219 In her cultural history: Barbara Ehrenreich, *Dancing in the Streets: A History of Collective Joy* (New York: Holt, 2006).

220 Zak proposes that oxytocin is a biological underpinning of trust: M. Kosfeld et al., "Oxytocin Increases Trust in Humans," *Nature* 435 (2005): 673–76.

220 the love of humanity: B. Campos, M. A. Logli, and D. Keltner, "Love of Humanity," unpublished manuscript.

221 the health of communities depends on trust and the love of humanity: Robert Sampson, "The Neighborhood Context of Well-Being," *Perspective in Biology and Medicine* 46 (2003): S53-S64.

221 children prove to be much more resilient in the wake of their parents' divorce when they feel a sense of connection: For several essays on more peaceful divorce, see Jason Marsh and Dacher Keltner, ed., "The 21st Century Family," *Greater Good* 4, no. 2 (2007).

222 loving relations get more important, and love all the sweeter: L. L. Carstensen and S. T. Charles, "Emotion in the Second Half of Life," *Current Directions in Psychological Science* 5 (1998): 144–49.

223 romantic love dips: Helen Fischer, *Why We Love* (New York: Owl, 2004).

223 I would ask them to read: Stephanie Coontz, *Marriage, a History* (New York: Viking, 2005).

224 to arrive at that magic ratio of five positive feelings for every toxic negative one that enables marriages: Gottman, *Why Marriages Succeed or Fail.*

224 It is a kelson of creation: *Song of Myself,* in Walt Whitman, *The Portable Walt Whitman,* ed. M. Van Doren (New York: Penguin, 1945), 36.

COMPASSION

225 historian Jonathan Glover documents many such "sympathy breakthroughs": Jonathan Glover, *Humanity* (New Haven, CT: Yale University Press, 1999).

226 As Charles Darwin developed his first account: Darwin, *Descent.*

227 Other influential thinkers in the Western canon: Nussbaum, *Upheavals of Thought.* For Nussbaum's comprehensive study of compassion in Western thought, see "Compassion: The Basic Social Emotion,"

Social Philosophy and Policy 13 (1996): 27–58. For another assessment of Western thought's approach to the emotions, see Oatley, *Emotions: A Brief History*.

227 "A feeling of sympathy is beautiful and amiable": Immanuel Kant, *Observations on the Feeling of the Beautiful and the Sublime*, trans. J. T. Goldthwait (Berkeley: University of California Press, 1960).

227 "If any civilization is to survive": Ayn Rand, "Faith and Force: Destroyers of the Modern World," *Philosophy: Who Needs It?* (Indianapolis: Bobbs-Merrill, 1982).

228 "A transvaluation of values": F. Nietzsche, *Beyond Good and Evil* (New York: NuVision Publications, 2007), 73.

228 "Hence a prince who wants to keep his authority must learn how not to be good": N. Machiavelli, *The Prince*, trans. G. Bull (New York: Penguin, 2003), chap. XV, 42.

228 In a series of controversial papers, physiological psychologist Steve Porges has made the case that the vagus nerve is the nerve of compassion: Stephen P. Porges, "Orienting in a Defensive World: Mammalian Modifications of our Evolutionary Heritage: A Polyvagal Theory," *Psychophysiology* 42 (1995): 301–17, and "Love: an Emergent Property of the Mammalian Autonomic Nervous System, *Psychoendocrinology* 23 (1998): 837–61.

229 people systematically sigh: June Gruber, Christopher Oveis, Jeffrey Newell, and Dacher Keltner, "Sighing and Compassion," unpublished manuscript.

230 Historians of science have rated Charles Darwin as off-the-charts in terms of kindness and warmth: Frank J. Sulloway, *Born to Rebel: Birth Order, Family Dynamics, and Creative Lives* (New York: Pantheon, 1996).

230 amid the noisy, loving spectacle of his ten children: Browne, *Charles Darwin: Voyaging*.

231 James was the progenitor: James, "What Is an Emotion?"

231 Walter Cannon, a student of William James's, was not so convinced by his advisor's provocative armchair musings: W. B. Cannon, "The James–Lange Theory of Emotion: A Critical Examination and an Alternative Theory," *American Journal of Psychology* 39 (1927): 106–24.

231 The blush, for example, peaks at about fifteen seconds: D. Shearn et al., "Facial Coloration and Temperature Responses in Blushing," *Psychophysiology* 27 (1990): 687–93.

232 when people are asked to guess whether their heart rate has increased or decreased: J. W. Pennebaker and T. A. Roberts, "Toward a His and Hers Theory of Emotion: Gender Differences in Visceral Perception," *Journal of Social and Clinical Psychology* 11 (1992): 199–212.

232 One-day-old infants: G. G. Martin and R. D. I. Clark, "Distress Crying in Infants: Species and Peer Specificity," *Developmental Psychology* 18 (1982): 3–9.

232 Many two-year-old children, upon seeing another cry: C. Zahn-Waxler, M. Radke-Yarrow, and R. A. King, "Child Rearing and Children's Prosocial Initiations Towards Victims of Distress," *Child Development* 50 (1979): 319–30; C. Zahn-Waxler et al., "Development of Concern for Others," *Developmental Psychology* 28 (1992): 126–36.

232 Pictures of sad faces: P. J. Whalen et al., "Masked Presentations of Emotional Facial Expressions Modulate Amygdala Activity without Explicit Knowledge," *Journal of Neuroscience* 18 (1998): 411–18.

232 So we asked first whether the exposure to harm would trigger activation in the vagus nerve: C. Oveis, E. J. Horberg, and D. Keltner, "Compassion and Pride as Moral Intuitions," unpublished manuscript.

233 These measures yield an index called respiratory sinus arrhythmia (RSA): G. G. Berntson, J. T. Cacioppo, and K. S. Quigley, "Respiratory Sinus Arrhythmia: Autonomic Origins, Physiological Mechanisms, and Psychophysiological Implications," *Psychophysiology* 30 (1993): 183–96. Cacioppo et al., "The Psychophysiology of Emotion," in *Handbook of Emotions,* 173–191.

234 In Singer's words, evolution has "bequeath(ed) humans with a sense of empathy—an ability to treat other people's interests": Peter Singer, *Expanding Circle* (New York: Farrar, Straus, & Giroux, 1981).

235 Take Paul Rusesabagina's remarkable heroism during the genocide of Rwanda: Philip Gourevitch, *We Wish to Inform You That Tomorrow We Will Be Killed with Our Families* (New York: Picador, 1998).

236 Within the social sciences, these courageous actions are readily attributed to selfish genes, to the desire to save kin, or to self-interest, pure and simple: Daniel Batson has taken the debate over altruism to a new level in his theoretical and empirical work. Rather than being misguided by either/or propositions (is there such a thing as altruism or not?), Batson proposes that most kind, pro-social behaviors are likely motivated by selfish and other-oriented motives. Perhaps more importantly, Batson has established a set of empirical guidelines for the documentation of more other-oriented, even selfless motives of altruistic behavior. C. D. Batson and L. L. Shaw, "Evidence for Altruism: Toward a Pluralism of Prosocial Motives," *Psychological Inquiry* 2 (1991): 107–22. Alfie Kohn has made a very similar point about the reluctance for people immersed in Western thought to attribute benevolent intentions to others, even when explaining the most altruistic kinds of action. Kohn, *The Brighter Side of Human Nature.*

236 altruistic action is a defense mechanism by which we ward off deeper, unflattering, anxiety-producing revelations about the self: G. Valiant, "Natural History of Male Psychological Health—V: The Relation of Choice of Ego Mechanisms of Defense to Adult Adjustment, *Archives of General Psychiatry* 42 (1976): 597–601.

236 In an essay on the sublime and the beautiful: Kant, *Observations on the Feeling of the Beautiful and Sublime*, 58.

236 To set the stage for his empirical studies: Batson and Shaw, "Evidence for Altruism." Nancy Eisenberg et al., "Relation of Sympathy and Distress to Prosocial Behavior."

237 In a first study: Batson et al., "Self-Reported Distress and Empathy and Egoistic versus Altruistic Motivation for Helping," *Journal of Personality and Social Psychology* 45 (1983): 706–18.

238 This age-old question motivated Batson's next study: J. Fultz et al., "Social Evaluation and the Empathy-Altruism Hypothesis," *Journal of Personality and Social Psychology* 50 (1983): 761–69.

239 just this kind of data: Nancy Eisenberg et al., "Relation of Sympathy and Personal Distress to Prosocial Behavior," and "Differentiation of Personal Distress and Sympathy in Children and Adults," *Developmental Psychology* 24 (1988): 766–75.

240 Our tendencies to experience specific emotions: Carol Z. Malatesta, "The Role of Emotions in the Development and Organization of Personality," in *Socioemotional Development: Nebraska Symposium on Motivation*, ed. Ross A. Thompson (Lincoln: University of Nebraska Press, 1990), 1–56.

240 the longitudinal studies of Harvard psychologist Jerome Kagan: J. Kagan, J. S. Reznick, and N. Snidman, "Biological Bases of Childhood Shyness," *Science* 240 (1988): 167–71; C. E. Schwartz et al., "Inhibited and Uninhibited Infants 'Grown Up': Adult Amygdalar Response to Novelty," *Science* 300 (2003): 1952–53.

241 Avshalom Caspi studied the adult lives of shy individuals: A. Caspi, G. Elder, and D. J. Bem, "Moving Away from the World: Life-Course Patterns of Shy Children," *Developmental Psychology* 24 (1988): 824–31.

241 In one study, Chris Oveis and I: C. Oveis et al., "Vagal Tone as a Biological Marker of Social Connection," unpublished manuscript.

242 Nancy Eisenberg has found that: N. Eisenberg et al., "The Relations of Children's Dispositional Empathy-Related Responding to their Emotionality, Regulation, and Social Functioning," *Developmental Psychology* 32 (1996): 195–209, and "The Role of Emotionality and Regulation in Children's Social Functioning: A Longitudinal Study," *Child Development* 66 (1995): 1360–84.

243 College students with higher resting vagal tone: Eisenberg et al., "The Relations of Emotionality and Regulation to Dispositional and Situational Empathy-Related Responding," *Journal of Personality and Social Psychology* 66 (1994): 776–97.

243 Following the loss of a married partner: M. F. O'Connor, J. J. B. Allen, and A. W. Waszniak, "Emotional Disclosure for Whom? A Study of Vagal Tone in Bereavement," *Biological Psychology* 68 (2005): 135–46.

243   And on the other end of the continuum: T. Beauchaine, "Vagal Tone, Development, and Gray's Motivational Theory: Toward an Integrated Model of Autonomic Nervous System Functioning in Psychopathology," *Development and Psychopathology* 13 (2001): 183–214.

243   an inspiration to James: In several places James expresses admiration for the open-spirited character of Walt Whitman. W. James, *The Varieties of Religious Experience*, (New York: Collier, 1902/1961), 88–89.

243   The vulnerability of our offspring: Hrdy, *Mother Nature*.

244   an instinct to care: Taylor, *The Tending Instinct*.

244   When parents look at pictures of their new babies: J. B. Nitschke et al., "Orbitofrontal Cortex Tracks Positive Mood in Mothers Viewing Pictures of Their Newborn Infants," *Neuroimage* 21, no. 2 (2004): 583–92.

244   the evocative power of the baby: Diane Berry and Leslie Zebrowtiz-McArthur have done terrific work on "neotony," or baby-faced appearance, which relates to all kinds of trustworthy perceptions and forgiving behaviors in others. D. Berry, and L. Z. McArthur, "Perceiving Character in Faces: The Impact of Age-Related Craniofacial Changes on Social Perception," *Psychological Bulletin* 100 (1986): 3–18.

244   the victory goes to the kind: Miller, *The Mating Mind*, chap. 9.

245   the largest study of mate preferences ever undertaken: D. M. Buss, "Sex Differences in Human Mate Preference: Evolutionary Hypothesis Tested in 37 Cultures," *Behavioral and Brain Sciences* 12 (1989): 1–49.

246   Darwin long ago surmised: Darwin, *Descent*, 130.

246   groups fare better when comprised of kind individuals: In support of this claim we have found that groups tend to select outgoing individuals who advance the interests of other group members as leaders. Group members also systematically identify unkind, Machiavellian types in gossip, to keep track of who poses threats to the interests of other group members. D. Keltner et al., "A Reciprocal Influence Model of Social Power: Emerging Principles and Lines of Inquiry," in *Advances in Experimental Social Psychology*, vol. 40, ed. M. Zanna (New York: Academic Press, 2008), 151–92.

246   In a study that explored this reasoning: C. Oveis et al., "Vagal Tone, Trust, and Generosity," unpublished manuscript.

247   what unites the ethics of the world's religions: Armstrong, *The Great Transformation*, chap. 7.

247   evolutionists would converge on a similar answer: Robert Trivers, Robert Frank, and Elliot Sober and David Sloan-Wilson have all argued how moral emotions like gratitude, compassion, and love bind individuals into cooperative bonds. Trivers, "The Evolution of Reciprocal Altruism," *Quarterly Review of Biology* 46 (1971): 35–57. See also Frank, *Passions Within Reason*; E. Sober and D. S. Wilson, *Unto Others*.

247   "My pedagogy is hard": Alice Miller, *For Your Own Good: The Roots*

of Violence in Child-Rearing, trans. Hildegarde Hannum and Hunter Hannum (London: Virago, 1987).

248 When Richie and Jon Kabat-Zinn and colleagues: R. J. Davidson et al., "Alterations in Brain and Immune Function Produced by Mindfulness Meditation," *Psychosomatic Medicine* 65 (2003): 564–70.

249 the kinds of environments that cultivate compassion: Nancy Eisenberg has written an excellent summary of the kinds of environmental factors that cultivate compassion. N. Eisenberg, "Empathy-Related Emotional Responses, Altruism, and their Socialization," in *Visions of Compassion*, 131–64.

249 Even visually presented concepts: M. Mikulincer et al., "Attachment, Caregiving, and Altruism: Boosting Attachment Security Increases Compassion and Helping," *Journal of Personality and Social Psychology* 89 (2005): 817–39.

249 In the words of the Dalai Lama: For a terrific statement about compassion, see His Holiness the Dalai Lama, *Ethics for the New Millennium*.

AWE

250 He wrote almost daily entries about these first experiences: John Muir, *My First Summer in the Sierras* (San Francisco, CA: Sierra Club Books, 1988).

252 finds in her research that adding trees and lawns to housing projects in Chicago: Frances Kuo has done several studies documenting that making urban settings greener brings about all sorts of benefits for individuals and communities. It is work that has important policy implications. Frances E. Kuo, "Coping with Poverty: Impacts of Environment and Attention in the Inner City," *Environment and Behavior* 33 (2001): 5–34.

252 "The way that can be spoken of:" Lao Tzu, *Tao Te Ching* (New York: Penguin Books).

252 evolutionists have recently begun to make the case: David Sloan Wilson and Elliot Sober are the most well-known advocates of the more general thesis that several adaptations enable more cohesive groups. One such example, Sloan Wilson argues, is religion, which builds stronger and more cohesive groups. They offer a variant of a group selectionist argument, arguing that these kinds of adaptations evolved to enable groups to out-compete other groups in group-to-group competition. The end result is that groups comprised of individuals with traits that give them an advantage vis-à-vis other groups will be more likely to survive and successfully replicate genes, and those traits will be selected for. This thesis is generating a good deal of controversy because it challenges the widespread assumption that natural selection operates only at the level of genes. I approach these group-related human adaptations from a different per-

spective, assuming that the human capacities that enable more cooperative groups—a sense of reverence for the group, art, dance, play—are selected for because they create conditions that enable less conflict and greater chances of survival and gene replication. From this perspective, group-related adaptations like awe influence survival and gene replication indirectly, through creating conditions more felicitous to natural and sexual selection. For a full treatment of these ideas, and a provocative account of multilevel selection theory, see Sober and Wilson, *Unto Others;* D. Sloan Wilson, *Darwin's Cathedral: Evolution, Religion, and the Nature of Society* (Chicago: University of Chicago Press, 2003).

253 traces back to Ralph Waldo Emerson: Emerson, "Nature," *Ralph Waldo Emerson, Selected Essays* (New York: Penguin, 1982), 39.

253 in particular Edmund Burke: Burke, *A Philosophical Inquiry into the Origin of Our Ideas of the Sublime and Beautiful* (Oxford: Oxford University Press, 1990).

253 Paul's conversion on the road to Damascus: "Letters to the Galatians," *The New American Bible*, ed. S. J. Hartdegen et al. (London: Thomas Nelson, 1987), 1320.

254 "Do works for Me": R. C. Zaehner, trans., *The Bhagavad-Gita* (London: Oxford University Press, 1969).

255 To bring some order to this cacophony of transcendence: D. Keltner and J. Haidt, "Approaching Awe, a Moral, Aesthetic, and Spiritual Emotion," *Cognition and Emotion* 17 (2003): 297–314.

256 Our experiences of powerful, charismatic humans: Our reasoning about the relationship between awe and power was profoundly influenced by Max Weber's writings on this topic. Weber, *Economy and Society: An Outline of Interpretive Sociology*, ed. Guenther Roth and Claus Wittich (Berkeley: University of California Press, 1978).

256 Aesthetic properties of the stimulus: For an excellent overview of the study of aesthetics, see M. C. Beardsley, *Aesthetics from Classical Greece to the Present* (New York: Macmillan, 1966).

256 Encounters with extraordinary virtue: Haidt and Keltner, "Appreciation of Beauty and Excellent (Awe, Wonder, Elevation)," 537–51.

257 The Greek philosopher Protagoras: Plato, *Protagoras and Meno*, trans. W. K. C. Guthrie (New York: Penguin, 1956).

257 In his beautifully distilled book: Paul Woodruff, *Reverence: Renewing a Forgotten Virtue* (New York: Oxford University Press, 2002).

258 small specks of time and matter in the vastness of the universe: Philosopher Thomas Nagel has long been interested in how shifts in perspective upon the self, wherein one looks upon the self from a detached, outside-the-self perspective, lead to states like the feeling of absurdity or awe. Nagel argues that in experiences like the sense of absurdity, we move from the absolutist demands of the inner and attached point of view,

where what is real, true, and right is only what I believe and see, to an alternative layer of meaning. We move, feeling light and open, to a perspective where we view our lives, again in Nagel's terms, from an outer, detached point of view. Thomas Nagel, *The View from Nowhere* (New York: Oxford University Press, 1986).

259 Evolutionists like David Sloan Wilson: Wilson, *Darwin's Cathedral.*

261 staging epiphanies: A. Joyce Nichols, *The Poetics of Epiphany: Nineteenth-Century Origins of the Modern Literary Movement* (Alabama: University of Alabama Press, 1987).

262 We found that goose bumps are fairly unique to awe: B. Campos et al., "Positive Emotion," unpublished manuscript.

263 Chris Oveis has found that the vagus nerve does indeed fire during the experience of elevation: C. Oveis, S. Sherman, J. Haidt, "Vagal Reactivity and Elevation," unpublished manuscript.

263 In one study, Lani Shiota and I had participants recall transformative experiences in nature: M. N. Shiota, D. Keltner, and A. Mossman, "The Nature of Awe: Elicitors, Appraisals, and Effects on Self-Concept," *Cognition and Emotion* (in press).

265 awe in the brain: E. Simon-Thomas, C. Oveis, and D. Keltner, "Positive Emotion in the Brain," unpublished manuscript.

266 The images of sensory pleasure: B. Knutson, J. C. Cooper, "Functional Magnetic Resonance Imaging of Reward Prediction," *Current Opinions in Neurology* 18, no. 4 (2005): 411–17. R. A. Depue and P. F. Collins, "Neurobiology of the Structure of Personality: Dopamine, Facilitation of Incentive Motivation, and Extraversion," *Behavioral and Brain Sciences* 22 (1999): 491–569; E. T. Rolls, "The Orbitofrontal Context and Reward," *Cerebral Cortex* 10 (2000): 284–94, and *The Brain and Emotion.*

267 the amygdala: LeDoux, *The Emotional Brain.*

267 known as the dorsal medial prefrontal cortex: R. J. Davidson, "What Does the Prefrontal Cortex 'Do' in Affect: Perspectives on Frontal EEG Asymmetry Research," *Biol Psychol* 67, nos. 1–2 (2004): 219–33; J. Mitchell, N. M. McCrae, and M. Banaji, "Dissociable Medial Prefrontal Contributions to Judgments of Similar and Dissimilar Others," *Neuron* 50 (2006): 655–63; K. N. Ochsner et al., "The Neural Correlates of Direct and Reflected Self-Knowledge," *Neuroimage* 28 (2005): 797–814.

267 This region lights up: Rolls, *The Brain and Emotion;* R. J. Davidson, "What Does the Prefrontal Cortex 'Do' in Affect."

268 For Charles Darwin: All quotes in this paragraph are from Browne, *Charles Darwin* I, *Voyaging.*

268 For cell biologist: U. Goodenough, *Sacred Depths of Nature* (New York: Oxford University Press, 1998).

268 "identification of ourselves with the beautiful which exists in thought, action, or person, not our own": Shelley, "A Defence of Poetry."

TEXT ACKNOWLEDGMENTS

ILLUSTRATION ACKNOWLEDGMENTS

CHAPTER 1
7 Dacher Keltner.
8 Dacher Keltner.
13 Dacher Keltner.

CHAPTER 2
18–21 Dacher Keltner.
23 Art Wolfe / Photo Researchers, Inc.
24 Corbis.
25 Dr. Lenny Krystal.
25 Dr. Lenny Krystal.
25 Dr. Lenny Krystal.
25 Dr. Lenny Krystal.
25 Dr. Lenny Krystal.
25 Dr. Lenny Krystal.
27 Courtesy of Paul Ekman.
33 Courtesy of Jeffery Cohn.

CHAPTER 3
37 Acro Images GmbH / Alamy.
37 James Kirkpatrick / Alamy.
38 Dr. Lenny Krystal.
38 Dr. Lenny Krystal.
38 Dr. Lenny Krystal.
38 Dr. Lenny Krystal.
38 Dr. Lenny Krystal.
40 Dr. Lenny Krystal.

40 Dr. Lenny Krystal.

43 Nisbett, R.E., Thomas D. Gilovich, and Dacher Keltner (2005). *Social Psychology*. New York: W. W. Norton.

49 Joshua Greene.

50 Dacher Keltner.

CHAPTER 4

58 David Gifford / Photo Researchers.

60 Frans Lanting / Corbis.

60 Xinhua / Landov.

69 Imperial War Museum, London.

70 Dacher Keltner.

CHAPTER 5

75 Eadweard Muybridge Collection / Kingston Museum / Photo Researchers.

77 Eadweard Muybridge Collection / Kingston Museum / Photo Researchers.

83 Dr. Lenny Krystal.

85 Darwin, C. (1872/1998). *The Expression of Emotions in Man and Animals*, 3rd ed. New York: Oxford University Press.

90 National Museum, Bangkok, Thailand, Giraudon / The Bridgeman Art Library.

90 Elliott & Fry / Getty Images.

90 Abbie Trayler-Smith / Panos Pictures.

93 Gazzaniga, Michael S., Richard B. Ivry, and George R. Mangun (2002). *Cognitive Neuroscience*, 3rd ed. New York: W. W. Norton.

CHAPTER 6

98 Gianni Dagli Orti / Corbis.

101 Clive Bromhall / OSF / Animals Animals - Earth.

101 Clive Bromhall / OSF / Animals Animals - Earth Scenes.

102 Interfoto USA / Sipa Press.

102 Tim Graham / Getty Images.

102 Chris Jackson / Getty Images.

106 Dr. Lenny Krystal.

106 Dr. Lenny Krystal.

106 Dr. Lenny Krystal.

106 Dr. Lenny Krystal.

107 Dacher Keltner.

109 Jan Steen, *Merry Company on a Terrace*, Metropolitan Museum of Art.

115 Ravenna Helson.
115 Ravenna Helson.

CHAPTER 7
127 Erich Lessing / Art Resource, NY.
129 Bo Veisland / Photo Researchers.

CHAPTER 8
149 Bettmann / Corbis.
152 Dacher Kelter.
156 Mary Evans Picture Library.
158–59 Dacher Keltner.
161 Dacher Keltner.

CHAPTER 9
174 Dacher Keltner.
179 Nucleus Medical Art, Inc. / Getty Images.
184 http://www.niaah.nih.gov/Recources/GraphicsGallery/Endo crineReproductiveSystem/Hypothalamic.html. Source: Adinoff, B., et al. Disturbances of the stress response: The role of the HPA axis during alcohol withdrawl and abstinence. *Alcohol Health & Research World* 22(1): 67–72, 1998.
189 Dacher Keltner.
190 Dacher Keltner.
192 Dacher Keltner.
193 Adrienne Gibson / Animals Animals — Earth Scenes.
193 Ingo Arndt / Foto Natura / Getty Images.
194 Juergen & Christine Sohns / Animals Animals — Earth Scenes.

CHAPTER 10
206 Lloyd Nielsen / OSF / Animals Animals — Earth Scenes.
209 Dr. Lenny Krystal.
209 Dr. Lenny Krystal.
214 MSOE Center For BioMolecular Modeling.
215 Dacher Keltner.
216 Dr. Lenny Krystal.
216 Dr. Lenny Krystal.
216 Dr. Lenny Krystal.
216 Dr. Lenny Krystal.
216 Courtesy of Frans Waal.
218 Kirk Cooper.

218 Kirk Cooper.
218 Kirk Cooper.
218 Kirk Cooper.
221 Dacher Keltner.
223 Dacher Keltner.

CHAPTER 11
229 Netter Illustration used with permission of Elsevier Inc. All rights reserved.
233 David Turnley / Corbis.
233 Justin Sullivan / Getty Images.
235 Dacher Keltner.
239 Dr. Lenny Krystal.
239 Dr. Lenny Krystal.
245 Dacher Keltner. Adapted from Buss, D. M. (1987). Mate selection criteria: An evolutionary perspective. In Crawford, C., M. Smith, and D. Krebs (eds.), *Sociobiology and Psychology: Ideas, Issues, and Applications* (pp. 335–51). Hillsdale, New Jersey: Lawrence Erlbaum.

CHAPTER 12
255 Dacher Keltner.
258 Dacher Keltner.
258 Dacher Keltner.
262 Dacher Keltner.
265 Dacher Keltner.
266 Dennis Gottlieb / Getty Images.
266 isisphoto / Newscom.
266 Mark Henley / Panos Pictures.
266 Blaine Harrington III / Corbis.

INDEX

Page numbers in *italics* refer to illustrations.